FREE-ELECTRON LASERS

FREE-ELECTRON LASERS

THOMAS C. MARSHALL

Professor of Applied Physics
Columbia University

MACMILLAN PUBLISHING COMPANY
A Division of Macmillan, Inc.
NEW YORK

Collier Macmillan Publishers
LONDON

For B. Marie Marshall

Macmillan Publishing Company
866 Third Avenue, New York, NY 10022

Collier Macmillan Canada, Inc.

Printed in the United States of America

printing number
1 2 3 4 5 6 7 8 9 10

Library of Congress Cataloging in Publication Data

Marshall, Thomas C.
  Free-electron lasers

  Bibliography: p.
  Includes index.
  1. Free-electron lasers.   I. Title.
TA1677.M37   1985   621.36′6   84-19370
ISBN 0-02-948620-3

Contents

Preface

It is never obvious when a technical book should be written about a new, developing research area. There is always an incentive to wait another year until things settle down, or the results are in on a crucial experiment. For those of us in universities, the decision is quite simple: either a book is written on a sabbatical, or it isn't written at all. This book is the consequence of such elementary planning.

One sabbatical unit ago, the author recalls working on the preliminaries of what is now known as the Raman FEL, as well as the Čerenkov FEL. The first reports of success at Stanford were beginning to circulate. Seven years has brought not only enormous progress, but enough time for the process of natural selection to filter out what is important and durable. This seems like a good time for an elementary book on Free Electron Lasers to appear, and in another seven years perhaps those of you who are becoming acquainted with the field for the first time will contribute to the text of a second edition.

A more serious reason to write now is that the larger scientific community needs to know more about Free Electron Lasers. There is a need for something between "news" and the professional paper, or for that matter, even the review paper. In addition, I believe the FEL community needs a book too, not necessarily to pull everything together in one neat treatise (which this is not), but rather to uncork the information in a less-specialized way, so that new people will be drawn into FEL research or applications. Our purpose is to introduce the free electron laser to the professional audience which can contribute, but which is not specialized enough at present to follow the technical literature with facility. It will become clear to the reader that the FEL has much to contribute to science in all areas of the spectrum, and that it is not interesting merely as an element in future military hardware.

This is not a monograph, or an encyclopedia of FEL wisdom, nor is it our objective that the reader will know enough to "build" a free electron laser. I have kept the book small enough so that it could be read in perhaps two or three evenings by someone with at least a bachelor's level training in physics or one of the related sciences. The mathematical development is on a simple level; when matters become complicated, physical arguments are made to avoid the appearance of a mathematical text. The reason for this is that I believe the reader is entitled to know what the facts are, even if there is not enough space to demonstrate these in a uniformly rigorous way. There is a bibliography which is referenced throughout the text, which may be helpful for more detailed study, and which also provides a running summary of who did what. Some portions of the book formed part of a course which I taught at Columbia in the spring of 1983, entitled "The Physics of High Energy Density."

The FEL field has benefitted enormously by yearly workshops, several of which I attended. We are all greatly indebted to the sponsors of these forums, as well as to the volumes of conference papers which have been most helpful in preparing this text. The author acknowledges the contributions of many of his professional colleagues for advice, suggestions, and particularly for the many illustrations and graphs which are sprinkled throughout the text. For those who have helped advance the understanding and implementation of the FEL, this may be an interesting "scrap-book".

CGS Gaussian units are used throughout the text.

List of Symbols

A	magnetic vector potential
a_s	$eE_s/k_s mc^2$
a_w	$eB_\perp/k_0 mc^2$
B_0	guiding magnetic field (axial)
$B_w, B_\perp$	undulator field representation, amplitude
B_s	scattered EM magnetic field
b_w	$eB_\perp/mc^2$
c	speed of light
d	gap spacing, diode spacing
e	electron charge
e_s	eE_s/mc^2
E_s	scattered EM electric field
f	filling factor
g, G	FEL gain (G, two-wave power gain; g, growth rate)
h	Planck's constant
I_b	electron beam current
J	current density
k_0	undulator wave number, $2\pi/l_0$
k_s	scattered wave wave number, $2\pi/\lambda_s$
k_i	idler wave number
l_0	undulator period
l_b	electron bunch length
L	undulator length
L_c	resonator length
L_b	distance between bunches
L_R	Rayleigh range
m	electron mass
M	number of EM wave bounces in resonator
n	electron density
N	number of undulator periods
p	an integer
P	canonical momentum or angular momentum
q	an integer
Q	resonator quality factor
r_0	classical electron radius, e^2/mc^2
r_b	electron beam radius
r_c	cyclotron radius of electron
R	geometrical radius [e.g., mirror, drift tube]
s	slip factor

t	time
T	period
u	velocity component
v	velocity component
$v_\perp$	quiver velocity amplitude; $v_{\perp 0}$ steady transverse velocity component
v_T	thermal velocity
V	potential
w_0	optical beam waist
x, y	transverse coordinates
z	longitudinal or axial coordinate

Greek Symbols

β	v/c
$\beta_\parallel$	motion along axis, $v_\parallel/c$
$\beta_\perp$	motion $\perp$ to axis, $v_\perp/c$
γ	$(1 - v^2/c)^{-1/2}$
β_r	resonance velocity parameter
γ_r	resonant energy factor
$\Delta\gamma/\gamma$	fractional change in electron energy
$\delta\gamma/\gamma$	spread in electron energies or momenta, normalized
Γ	spatial growth rate
ε	emittance
η	efficiency
θ	angle with respect to FEL axis
Θ	dimensionless FEL parameters, Chapter 4
λ	wavelength
λ_D	Debye wavelength
$\nu_{L,i}$	loss rate (EM or ES wave energy)
ν	Budker's parameter
μ	refractive index
τ	time parameter
ϕ	phase angle of optical wave
ψ	phase angle of electron motion in undulator
ω	radian frequency
ω_0	undulator frequency, $k_0\beta c$, or $\omega_0' = \gamma k_0\beta c$.
ω_{p0}	plasma frequency, $(4\pi n e^2/m)^{1/2}$
ω_b	bounce frequency
ω_p	invariant plasma frequency, $(4\pi n e^2/\gamma m)^{1/2}$
Ω_L	reciprocal bounce distance
Ω_0	cyclotron frequency in guiding field, eB_0/mc
$\Omega_\perp$	cyclotron frequency of undulator field, $eB_\perp/mc$

1

Introducing Free-Electron Lasers

1.1 *Definitions and Comparisons*

There has been increasing research devoted to generating coherent radiation in the centimeter to visible regions of the spectrum through the use of relativistic electron beams. This interest can be understood as the continuation of a development which began with the magnetron and continued with such inventions as the klystron and the traveling-wave tube; progress was interrupted during the period when the discovery of various lasers was most active (1960–1970), but it has now resumed. The past decade has seen the rapid development of the electron cyclotron maser [λ = 1 cm to 1 mm] and more recently, the free-electron laser [λ = 1 mm to $< \frac{1}{2}$ μm]. The purpose of this book is to provide a simple introduction to, and survey of the physics of free-electron lasers.

The free-electron laser (referred to henceforward by its acronym, "FEL") is a device which amplifies short-wavelength radiation by stimulated emission, using a beam of relativistic electrons. We shall begin this book by clarifying some of the words used in this definition.

Strictly speaking, in the FEL the electron is not "free," since it is under the influence of magnetic forces which cause it to radiate, but it is "free" in the sense that it is not bound into an atom as in the case of the conventional laser. The FEL radiation is usually caused by passing the electron down a magnetic device known as an "undulator" or "wiggler," in which the electron is forced to execute a periodic oscillatory trajectory in space. The undulator may be a helical field, produced by a bifilar helical winding, which will guide the electron along a nearly helical orbit, or it may be a "linearly polarized" field made by a set of alternating-polarity magnets. On the other hand, the undulator can be an electrostatic device, or perhaps even a high-intensity light wave. The FEL usually operates in a vacuum, but if a dielectric is nearby, the Čerenkov effect (electron moving through a medium superluminously) may become important. Close relations of the FEL are devices such as the orotron which use the Smith-Purcell effect (where the electron radiates while passing over a grating) or devices which involve spiraling electrons and no undulator (which are related to the gyrotron, or cyclotron maser]. In all FEL devices one is not dealing merely with the emission of spontaneous or noise radiation, but also with the process of induced emission, which releases appreciable amounts of power.

Figure 1.1 is a schematic of the canonical FEL, operated by Madey and co-workers at Stanford University in 1977, which was configured as an oscillator with a pair of mirrors constituting the Fabry-Perot resonator. The undulator—N periods long, each period l_0 cm, so that the total undulator length is $L = Nl_0$ cm—is located between the mirrors. For particular choices of beam energy, cavity length, etc., the device will generate coherent radiation at a wavelength which depends on the electron energy and undulator period according to the approximate relationship

$$\lambda_s = \frac{l_0}{2\gamma_\parallel^2} \tag{1.1}$$

where λ_s is the radiated wavelength and $\gamma_\parallel$, the relativistic factor, is $(1 - v_\parallel^2/c^2)^{-1/2}$ where $v_\parallel$ is the electron velocity along the axis of the FEL. The total $\gamma = (1 - v^2/c)^{-1/2}$ is related to the particle kinetic energy W by $(0.51 + W)/0.51$, where the units of W are MV. Equation (1.1) is the FEL relation, and is sometimes referred to as the "resonance relationship"; if a device is an FEL, the frequency radiated will usually scale as $2\gamma_\parallel^2$, which is also roughly equivalent to $(1 - v_\parallel/c)^{-1}$. Since the undulator period is generally of order 1 cm, the radiated wavelength will be in the infrared if $\gamma_\parallel$

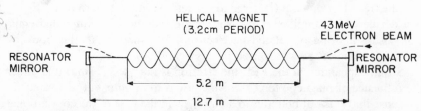

HELICAL MAGNET
(3.2cm PERIOD)

43 MeV
ELECTRON BEAM

RESONATOR
MIRROR

RESONATOR
MIRROR

5.2 m

12.7 m

Figure 1.1 FEL oscillator of Deacon et al. (1977). In addition to the undulator, a guiding field was used to route the beam around the mirrors. © 1977 APS.

is about 10, i.e., the electron energy is about 5–10 MV. Therefore we are dealing with relativistic situations.

The word "laser" has been associated with devices using certain excited energy levels in atoms or molecules, but it is also generally understood to apply to any coherent source of radiation at short, or optical, wavelength. The border between "short" and "long" wavelengths can be taken as the zone where optical or quasi-optical techniques are used, rather than guided-wave or resonant structures appropriate to microwave or maser devices. In round numbers; the borderline wavelength is about 1 mm (1000 μm); this will be the longer-wavelength limit of FEL technology discussed in this book. The span of FEL wavelengths extends beyond the visible, perhaps into the soft X-ray region. The range of tunability is limited in any FEL by the flexibility of the electron accelerator, but one can expect that an FEL should be able to explore about a decade in wavelength without a major design alteration. In Fig. 1.2 we see how various accelerators can be used for FELs operating in different spectral domains, and one appreciates how FEL technology has benefited from the preceding fifty-year period of accelerator development.

Interactions in the FEL are nonlocal in character. In a conventional laser, given uniform density and pumping, the gain in the medium does not depend on the position of the atom. In an FEL the electrons convect along the system, are bunched and perhaps trapped into waves, or otherwise interact in a way that depends on upstream conditions. The gain also depends upon the direction of the amplified electromagnetic (EM) wave.

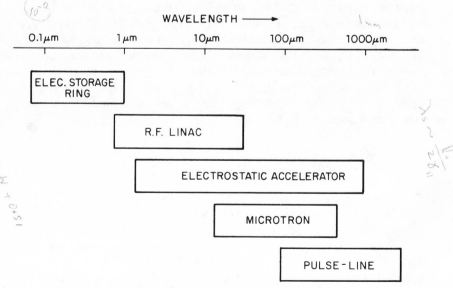

Figure 1.2 Range of FEL wavelength and accelerator technology.

The wavelength of the FEL depends on external parameters—the undulator periodicity and the beam energy—rather than on a fixed, internal transition within an atom. The feature of tunability therefore means the FEL will not be suitable as a frequency standard. There is a degree of coherence in the FEL system having to do with the fact that the electron experiences N periods of oscillatory motion in the undulator. There is a further improvement of coherence when the FEL is made to oscillate by applying optical feedback in a Fabry-Perot resonator. Nevertheless, the coherence of FEL radiation cannot match that of the gas laser.

The FEL can be described using the equations and laws of classical physics. The wavelengths are such that $\hbar\omega_s/mc^2 \ll 1$, and we shall be dealing always with situations where there are many photons per EM mode. These conditions also obtain for ordinary lasers. However, in the latter, radiation arises from a quantum-mechanical process, whereas in an FEL the radiation process can be understood classically. Unlike an excited atom, an electron does not decay; also, the electron is capable of multiphoton transitions even when the excitation is weak. Quantum-mechanical descriptions of the FEL have not been essential up to this time, given that the number of photons and electrons is large, but it is customary to draw upon such models whenever it clarifies understanding. Extension of FEL operation into the UV or soft X-ray region may change this situation somewhat.

Use of the word "laser" must not mislead the reader into believing that the similarities are so important that nothing is particularly new, and the FEL is just another variety of laser which uses a peculiar state of matter. There are many important differences, even though the physics of stimulated emission (or, for the FEL, stimulated scattering) is common to both. It has taken several years to refine FEL theory and provide experimental backing to the models which form the basis of our thinking. In this book our intention is not to write down a set of comprehensive equations and then derive specific results for each case; this would be possible only by extensive numerical simulation. Instead we shall build up the conceptual basis of the subject by starting with the simplest models and adding refinements sequentially.

1.2   Technology

The FEL provides a link between the technology of quantum electronics—high-intensity coherent light—and the technology of energetic electron beams and accelerators. The appearance of the FEL is timely, occurring when the physics of lasers and accelerators is highly developed. There is accordingly some 50 years of accelerator technology which can be used for generating coherent radiation, together with perhaps 25 years of advanced optical technology which can be used for accelerating charged

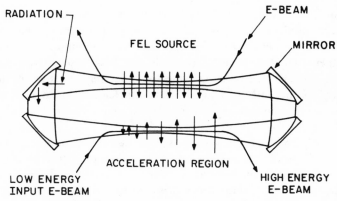

Figure 1.3   Use of an FEL as an accelerator (after Sprangle [1981]). © 1981 IEEE.

particles. We shall see that an FEL operated in reverse (stimulated absorption) becomes a device for converting the energy of coherent light into kinetic energy of an electron. This suggests that the FEL and its associated accelerator constitute a very well-integrated system: one accelerator can stimulate coherent radiation in a given undulator, and that radiation can itself be used to accelerate another electron stream (Fig. 1.3). The energy of the "used" electrons can be recovered in a deaccelerator system to enhance the overall system efficiency.

Theory in such areas as accelerator physics, plasma physics, and quantum electronics has already been tapped to provide a basic theoretical understanding of the FEL. In certain accelerators the physics of energy transfer from rf fields to electron motion is closely related to the physics by which an electron releases kinetic energy to an EM wave in the FEL undulator.

Recent experiments have shown that energy recovery from high-energy electron beams is not only practical but very efficient—on the order of 95–99% effective. This is an important consideration in calculating the net efficiency of the FEL system. To this date FELs have not attained the efficiency of well-engineered microwave tubes [say > 50%]. Part of the reason has to do with the immediate need to test the basic principles in existing accelerator facilities which are not optimized for the FEL application. Fundamental relationships limit the efficiency (in converting electron energy to radiation) of a simple FEL configuration to a few percent, with unfavorable scaling to shorter wavelengths. However, there are several ways to circumvent this limitation if one is willing to complicate the system or accept certain constraints. Furthermore, even an FEL efficiency of only 2%, when combined with a 95% energy recovery in the accelerator, will give a net efficiency $\approx 25\%$. Of course one must bear in mind that obtaining efficiency of this order will increase the investment required and will extract a certain toll in system flexibility. For example, a high-efficiency, variable-parameter undulator FEL constrains the amplified wavelength and the input power level; the net gain may also be less than optimum. Thus one must keep other factors in mind, and recall that most lasers have rather low efficiency but are still very desirable for other reasons. For instance, the tunability of the FEL is a highly desirable feature which is particularly valuable to chemists. Only a few lasers—e.g., the excimers and the CO_2—have efficiency comparable with the FEL.

The FEL operates in a vacuum where the only medium is a stream of "free" electrons at megavolt energy, having density in the range of 10^9 to 10^{12} cm^{-3}. Apart from exterior elements, there is no dielectric through which the amplified radiation need pass. At high intensity, the FEL should not experience those faults associated with dielectric failure. There are, nevertheless, certain effects—such as the sideband instability and self-focusing—which can compromise FEL performance at very high intensity.

The radiation hazard from the high-energy electron beam usually requires that the FEL be located in a specially prepared area. This problem clearly does not add to the attractiveness of the FEL, and the decision to build an FEL facility must be undertaken carefully. Adding an energy recovery system or using a closed-ring accelerator will enormously reduce the fraction of electron energy that is converted into ionizing radiation.

The FEL is a rather complicated device, and furthermore it has arrived when there is considerable competition from well-established conventional lasers. It must justify itself through its unique ability to satisfy needs not met by other sources. We shall list a few of these in Section 1.4.

1.3 Components

The most important physical component of an FEL is the accelerator. The energy of the electron beam produced by the accelerator will be in the relativistic range for electrons, $> mc^2 = 510$ kV, and it may be as high as 1000 MV (Fig. 1.4). The current from the accelerator will be in pulsed form as a rule; the pulses might last as long as several hundred microseconds or as little as a few picoseconds. The repetition rate will range from single-shot operation to perhaps 1000 Hz in continuous operation. The electrons are accelerated in a diode structure, or *gun*, which includes either a hot or a cold cathode, together with focusing elements and/or a guide magnetic field. Partial, if not complete, acceleration will occur in the diode.

Another parameter which characterizes the beam is its intrinsic parallel momentum spread (a completely "cold" beam will have only one value of parallel, or axial, momentum for a given energy). Ideally, all the electrons emerging from the accelerator would have the same energy, but owing to the details of the accelerator geometry and transport system, the electrons

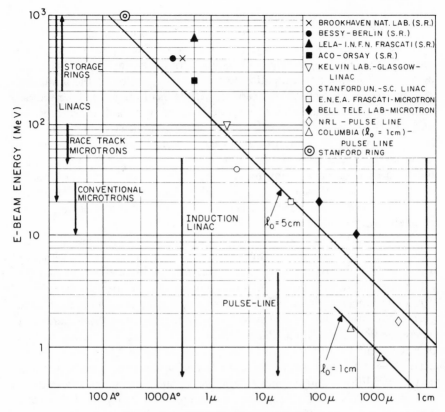

Figure 1.4 FEL experimental facilities.

would still have a distribution of momentum in the transverse and axial directions. This effect is described by a parameter known as the beam emittance, which characterizes the *brightness* of the beam. FELs require bright electron beams with small parallel momentum spread. This is not surprising when we recall that the output of a conventional laser is also "bright" (viz., in addition to high power output, the light beam has very little divergence and can therefore be focused to a diffraction-limited spot). We shall focus upon the accelerator qualities more in Chapters 5, 7, and 8.

A device is needed which will transform the linear, nonradiating motion of the electrons to a state which will allow coupling to the EM field. Since the EM field of a light beam moving parallel to the axial motion of the electrons is transverse, in order for the electron to exchange energy with the EM field it must have a component of motion transverse to the axis. Then the electron can do work on the field via the power term, $j \cdot E$. This is done by bending the path of the electron in a periodic, transverse, magnetostatic field which is provided by the undulator. In this case one would refer to the radiation produced as "magnetobremsstrahlung." There are many refinements involved in accurately calculating the motion of the electron in the undulator, simple as the situation may seem at first sight, and we shall return to this question in detail in Chapter 5. In the meantime, in Chapter 2, only a very simple approximation to the electron motion in the undulator will be used (it neglects off-axis motion, self-fields of the beam, longitudinal components of the undulator field, etc.). There are many other types of undulators too: periodic electrostatic elements, EM waves of high intensity, and perhaps even an electrostatic wave which could be developed by injecting the electron beam into an appropriate plasma, or crystal. Most of these variants are less convenient and reliable than the magnetostatic undulator, and they will not be discussed in detail. Most undulators either are made up of helical current windings, or are linear arrays of permanent magnetic dipoles.

A typical value of the undulator period is about 3 cm, and a typical undulator field amplitude is in the range of 1 kG. This field can be set up by a pulsed or DC current [the latter possibly in a superconductor], or on the other hand one can assemble an undulator out of permanent-magnet "building blocks" (such as samarium-cobalt). Simple undulators have fixed period and amplitude, except for a short zone near the entrance and exit where the field is increased adiabatically (*taper*) so that the electron motion proceeds smoothly into the new region. It has, however, become possible to design the undulator so that the FEL performance is improved under certain conditions, and this has resulted in a stable of nonuniform undulators where the period, amplitude, or configuration changes adiabatically along the structure. The successful operation of FELs with nonuniform undulators has greatly expanded the possible uses of this instrument and has provided additional confidence in theoretical understanding.

As we shall demonstrate later, the undulator permits the FEL to be not only an electron decelerator but also an electron accelerator. This is also possible in, say, the linear accelerator, where acceleration or deceleration is accomplished by the longitudinal component of the rf electric field in a set of cavities, or a loaded transmission line. However, the interaction in the linac is always at the frequency of the rf source, while in the FEL the interaction takes place at the valuable double-Doppler-shifted frequency. The undulator is the coupling device between a light wave and the relativistic stream of electrons.

Another important FEL component is the set of resonator mirrors. There is some tendency to regard this as an already solved problem from traditional laser optics, but that is not so. Improvement of mirrors is important to all areas of laser physics, and this is especially so for the FEL, particularly with regard to the following topics. First, there is the matter of mirror reflectance, which is crucial for low-gain FELs, especially those in the visible and UV. Stable, broadband coatings are needed. The optimum reflectance is a classified research area, but $R \approx 0.9995$ is possible using several [e.g., ~ 20] interference films in the visible, and $R > 50\%$ can now be obtained at $\lambda \sim 100$ Å. "Tayloring" of the reflectance function may be useful in controlling the sideband instability, which is believed to afflict FELs operating at high power (Chapter 3). Second, when the FEL is operated with a storage-ring electron device—where the electron energy may be 100–200 MV—substantial UV synchrotron radiation will deteriorate a typical high-reflectance coated mirror in short order. Development of high-reflectance mirrors stable against UV radiation is a precondition for FEL progress in this area. Third, power dissipation in the mirror may destroy the surface in short-wavelength, high-power applications. For example, an FEL output of 10 MW through a 1%-transmitting mirror would not be regarded as unreasonably high; yet the resonator power would be ~ 1 GW. The mirror substrate must dissipate a small fraction of this contained power without causing progressive surface deterioration. Although a 1-GW pulse in the cavity does not present a difficult problem with current technology, the power available from certain accelerators is very high, and this will become more of an issue as the FEL realizes its potential. Finally, long-wavelength FELs may benefit from feedback provided by distributed reflecting elements ("distributed feedback"), a technology still in the research phase [106].

The interaction of the electron beam with the electromagnetic modes in the optical resonator is an area where there is considerable contact with optical electronics [14]. Both transverse and longitudinal modes are set up between the resonator mirrors as a result of multiple reflection of the light rays, in a way originally described by Fox and Li. Losses in this Fabry-Perot resonator depend on diffraction loss around the mirrors, dissipation on the mirror surface, and coupling (via a hole, unstable resonator, or partly transmitting mirror coating). The electron beam will oscillate in those modes where the interaction is strong (i.e., there is good overlap of the electron and optical structure) and the losses are low. As the FEL linewidth is broad by atomic-laser standards, many resonator modes can be excited unless filters are introduced. In certain FELs, the electron-beam structure is a pulse about 1 mm in diameter and only a few millimeters long, with only one electron pulse in the cavity at a time. In this case, there is a tendency for the light wave to move off the more slowly traveling electron pulse, an effect referred to as *laser lethargy*. The output is a series of narrow mode-locked pulses.

1.4 Applications

A cursory inspection of the spectrum (Fig. 1.5) shows that the FEL would be most novel in the spectral range where coherent sources are few, e.g., the submillimeter and the ultraviolet. It is still too early to compare the FEL system with more highly developed laser systems, since very little investment has been made in specialized power-supply and accelerator apparatus. However, an FEL in the submillimeter domain need occupy no more space than a typical submillimeter molecular laser. Even in the near IR and visible spectral regions, the FEL may be competitive with conventional lasers in certain applications where tunability, power, or efficiency is important.

In the chemical "fingerprint" zone (200–4000 cm^{-1}) a tuneable source would make vibrational spectroscopy of absorbing molecules more convenient. FEL pulses can be shorter than molecular relaxation times; hence bond-selective chemical processes can be precisely controlled. An efficient FEL would contribute to the understanding as well as the commercial exploitation of laser-induced chemical chain reactions, where a photodissociation of one molecule will catalyze the formation of large numbers of other molecules [58]. In the far IR [~ 100 μm], where the photon energy is roughly kT, the FEL could be used to study shallow van der Waals states. Wide tunability and high efficiency make the FEL attractive for laser isotope separation and enrichment.

Another highly useful property of certain FELs is the short-pulsed mode of operation (when the FEL is associated with a microtron, rf linac, or storage ring). The short pulses (Fig. 1.6), a few picoseconds long, are well matched for the study of various excitations (Fig. 1.7) [167]. Sources of

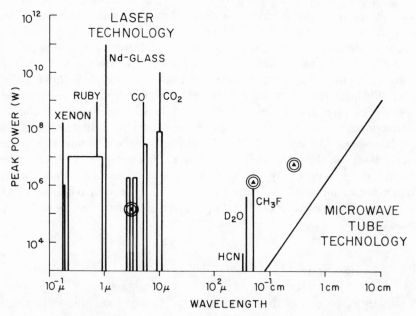

Figure 1.5 High-power coherent sources; FELs indicated by double circles.

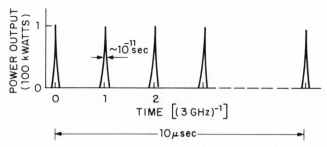

Figure 1.6 Schematic of FEL power pulse from an rf linac. The spikes are the micropulses, which together form a macropulse lasting several microseconds, which is itself repeated at several hertz.

far-IR and submillimeter power had—until the invention of the FEL—been weak and scattered, while detectors have been expensive or insensitive. There is, however, a wealth of important data which can be extracted with a far IR FEL facility (Fig. 1.8) [167] dedicated to solid-state studies, such as the one constructed at Bell Laboratories. Nonlinear spectroscopy and transient studies should be the principal beneficiaries of a far-IR pulsed FEL. The former include stimulated-emission phenomena, inelastic scattering from electronic excitations, optical pumping, etc. The latter involve energy transfer mechanisms in molecular, solid, and liquid systems, including the relaxation of the hot electron-hole gas in semiconductors. Short pulses of far-IR radiation are a good probe for local electrical conductivity,

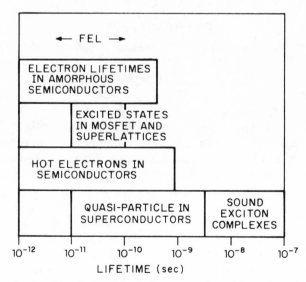

Figure 1.7 Applications for FEL radiation source in solid studies [after Shaw and Patel (1980)].

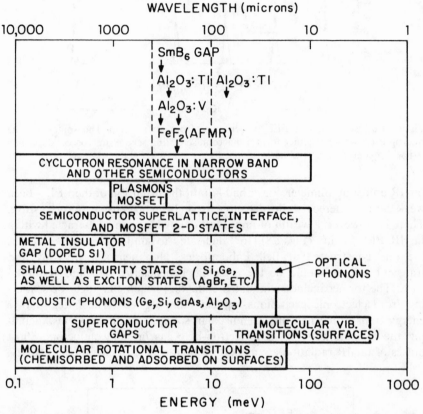

Figure 1.8 Applications for FEL radiation source in solid state studies [after Shaw and Patel (1980)].

while visible FEL pulses will generate free carriers. Other possibilities include the investigation of quasiparticle scattering times in excited super-conductors, studies of phonon propagation and interaction, and the excitation and relaxation of the two-dimensional electron gas in MOSFETs [10].

Surely one of the most important and timely problems in semiconductor physics which can be understood with the aid of a submillimeter FEL is the question of hot electrons. The increasing miniaturization of semiconductor elements tends to promote nonthermal electron distributions, because the constant energy gap [~ 1 V] results in high electric fields. As the effects of hot electrons are not well known, an FEL submillimeter source would be useful to create as well as to probe the transient behavior of these carriers [58].

The high average-power capability of the FEL at $\lambda \approx \frac{1}{2}$ mm suggests an application to the heating of plasmas contained in a strong magnetic field by electron cyclotron resonance, or to the heating or diagnosis of high-β plasmas ($\beta = 8\pi n k T / B^2$) for various fusion-energy applications. In laser-induced nuclear fusion, there is a need for high power, short-pulse radiation at $\approx \frac{1}{3}$ μm, with overall efficiency of several percent, and delivery of ~ 1 MJ on target. FEL sources may be able to contribute to this program in the future [145].

Remote sensing in the upper atmosphere (100–500 km) using resonance fluorescence and lidar for molecular species is another area of diagnostic interest. The FEL should be useful in the development of high-resolution radars.

To this date FELs have not operated as oscillators in the UV region; there are many applications for tunable intense sources of UV, particularly in solid-state science, but until the technical difficulties are overcome, we shall have to be content with the spontaneous radiation from electrons passing through the undulator in high-energy storage-ring facilities—the latter attached to a synchrotron-radiation project. Improvements is mirror technology and the development of special accelerator facilities now underway may result in an operating FEL in the UV before 1990. Development of an X-ray FEL would be invaluable for such purposes as the fabrication of high-resolution optics by X-ray interferometry and holography.

The FEL can contribute to laser surgery and photoradiation medicine. For surgery, its small spot size and tunability mean that a wavelength-dependent effect on a particular tissue can be obtained. In photoradiative medicine, dyes are injected into tissue and then activated at a specific wavelength. The activation of a dye could release free oxygen, thereby removing a cell by nonsurgical means. Dyes may be attached to antibodies which are released at specific locations by the laser light. With a tunable laser, more dyes become available for this type of work. A relatively low-power FEL might therefore find application in a hospital.

Accelerators used in high-energy physics have become exceedingly large, and a search is underway to determine if higher accelerating fields may be used to reduce accelerator size while increasing particle energy. It is known that focused high-intensity laser fields can produce transverse electric fields $\sim 10^9$ V/cm; this could be used in conjunction with an FEL accelerator (e.g., Fig. 1.3) to accelerate a beam of positrons or electrons moving along an undulator. As we shall learn in Chapter 3, changing particle energy can be accommodated if the period of the undulator field is changed in step. Starting with a period ~ 10 cm, and increasing to several meters, electrons could be accelerated to > 100 GV in several kilometers. The difficulty is to maintain an intense focused laser beam over this great distance: Pellegrini [154] has proposed to do this using an optical waveguide. If this proposal is successful, the FEL may repay its debt to accelerator physics.

1.5　*Operating Regimes and Classifications*

The FEL will operate in several distinct regimes where the physical principles are quite different, and an extensive nomenclature has evolved to characterize the important parameters. We give a brief summary here.

If the electron beam current is low, the beam energy is high ($\sim$ 20 MV for example), and the wavelength is short (IR), then we are in the regime variously described as "Compton," "two-wave," or "interference" FEL, or the "individual-particle" interaction. In this domain there is a very close analogy between the FEL and the rf accelerator: a gain (loss) of particle energy corresponds to a loss (gain) of EM field energy. Optimum gain of the FEL depends on a specific choice of beam energy and undulator length: hence, another term for this regime is "finite undulator length." The notion from laser physics that more medium means more gain and power does not necessarily apply here. Furthermore, no gain occurs when the optical beam reflects off the mirror at the end of the FEL and travels antiparallel to the streaming motion of the electrons.

FELs having exponentially growing waves, and which resemble conventional pumped lasers more closely, are operated in the regime of long wavelength ($\lambda > 100 \ \mu$m), low energy (typically less than 5 MV), and dense beam ($j \gtrsim 1000$ A/cm^2); these are compact, high gain devices. If the beam is *cold* (viz., the electron momentum spread is sufficiently small) and the pump field from the undulator is weak, we have the case of the *Raman* FEL. As the undulator pump field is increased, the gain coefficient increases and one enters the *strong-pump* regime of optimum gain and efficiency (this is sometimes referred to as the "oscillating two-stream instability"). On the other hand, if the beam is too warm, the gain and power fall but the signal still grows exponentially along the undulator, in what is called the *warm-beam Compton* regime. The gain falls because the Raman FEL is a three-wave parametric device (pump, signal, idler) in which the idler wave is a plasma or space-charge wave, and when the idler is damped by a collisionless warm-beam effect (Landau damping) we revert to lower gain. [All FELs require a sufficiently cold electron beam.]

A rather simple demarcation can be drawn between the single-particle FELs and those where collective effects, which give rise to exponential gain, are significant. The FEL operates in the collective [many-particle] regime if the system is sufficiently long and dense that it is several plasma oscillations in size. This sets an upper limit on the beam energy and a lower limit on the wavelength. Another very important effect is the interplay between the ponderomotive force (which bunches the beam electrons) and the repulsive space-charge forces. The ponderomotive force is governed by the amplitudes of the undulator and the signal, while the space-charge forces depend upon the beam density and energy. In some instances we shall find that the FEL gain is enhanced, and in other cases that it is diminished, by this competition.

As with atomic lasers, FELs can operate as amplifiers for coherent radiation, as oscillators (using a cavity resonator with mirrors), or as amplifiers of local noise. The last case we characterize as *superfluorescent*, after the optical terminology. When the gain is very high, a superfluorescent FEL can be a powerful source of partially coherent radiation.

It has been customary to name various electron devices with the suffix "-tron" (ubitron, scattron, oratron, etc). Apparently there is no methodology to this tradition. There is still a tendency to add to this nomenclature in the microwave or millimeter-wave region where the wave speed is dependent upon the geometrical properties of the structure. In the free-electron laser, however, the electromagnetic wave speed is c, to a very good approximation.

1.6 *Historical Survey, 1950–1980*

The idea of the stimulated Thomson or Compton effect dates back as far as 1933, to a publication of Kapitza and Dirac [104], but found no potential application until the 1950s, when much effort was devoted to pushing the limits of electron-tube technology to shorter wavelength. The scaling of these devices with decreasing wavelength is unfavorable with respect to precision, tolerances, and power; furthermore, investigations were limited to the low electron energy regime (< 10 kV) as a rule. Results were not impressive, and with the invention of the laser in 1960, research interest shifted into the newer area of quantum electronics. Noteworthy, however, is the effort that Motz [140, 141] initiated at Stanford on the spontaneous radiation from an electron beam moving through an undulator at relativistic energy. The spectrum and power obtained were compared with the predictions of single-electron radiation formulas of classical electrodynamics. Experiments [141] were done in both the optical and millimeter spectral regimes. In 1959, Motz and Nakamura [202] showed that external waves could be amplified in a waveguide structure.

As an outgrowth of Motz's research, Phillips [155] developed a microwave device referred to as the *ubitron* (we would now classify the ubitron as a weakly relativistic FEL device operating in the strong-pump regime). Results were good [$V = 150$ kV, $P > 100$ kW, efficiency $\sim 10\%$] at 10-cm wavelength, but other devices also have favorable output at this wavelength, and the development of the ubitron as a microwave source was temporarily eclipsed by the gyrotron.

Interest in Motz's work continued at Stanford, and in 1968 Pantell et al. [148] proposed a device similar to the one used by Motz, but it was now equipped with a pair of mirrors and a microwave undulator. About 1970,

John Madey, with the assistance of Professor H. A. Schwettman, began research on what he referred to as "a free electron laser" [125]. The first to comment upon the possibilities of an FEL in connection with accelerator technology was Palmer [147]. The first relativistic stimulated-scattering experiments in the two-wave region were carried out at Stanford, and in 1976, Elias et al. [65] utilized a low-current electron beam in a linear accelerator [$I = 70$ mA, $V = 24$ MV] to achieve 7% amplification at 10.6 μm using a helical undulator field of 2.4 kG. Later a similar experiment was performed in the oscillator mode by Deacon et al. [54], achieving a peak power of 7 kW at 3.4 μm utilizing the same pump configuration with a more intense and higher energy electron beam [$I = 2.6$ A, $V = 43$ MV]. The 7-kW power output represented 0.01% of the beam energy. Since amplification is due to an interference effect, the gain is strongly dependent on the finite undulator length. The initial analyses of the FEL were quantum-mechanical [125, 187, 126]. However, subsequent research showed that a classical treatment of the stimulated scattering process is satisfactory [205, 46].

Meanwhile, using generators of intense relativistic electron beams (1–10 kA/cm^2, 1–2 MV), Nation [142] and Friedman and Herndon [78] studied radiation emitted by slow-wave structures or rippled-magnetic-field elements. An investigation of powerful submillimeter radiation at NRL [87, 88] was attributed to stimulated scattering, using an interpretation by Sprangle et al. [182]. The first series of experiments using a static-magnetic-field undulator for excitation of the Raman scattering process was conducted at Columbia University. Efthimion and Schlesinger [64], generating high power at wavelength ~ 0.3–6.0 cm, showed that the physical mechanism involved the coupling of waveguide modes to the negative-energy cyclotron and space-charge waves as idlers [129]. Marshall et al. [131], reconfiguring the experiment and operating with a more intense pump field, succeeded in producing several megawatts of power in the 1–3-mm range, and by using an electromagnetic undulator that permitted the variation of pump field, established the linear dependence of growth rate on pump amplitude, implying that the interaction was in the Raman regime [182]. Detailed spectroscopic studies of the radiation in the millimeter band were carried out by Gilgenbach et al. [84]. McDermott et al. [132] reported on the realization of a collective Raman FEL configured as an oscillator. This device, involving a cooperative Columbia-NRL effort, used a quasi-optical cavity to permit optical feedback and a special high-magnetic-field zone near the cathode to reduce beam momentum spread. Laser output of 1 MW at 400 μm and line narrowing to 2% were observed; subsequent experiments [30] confirmed the FEL wavelength scaling relationship. By redesigning the diode and electron beam [151], the NRL group was able to demonstrate an efficiency in the 7% range for the Raman FEL. In a different application of the FEL, Walsh [195] conducted a series of experiments to demonstrate the feasibility of combining the stimulated Čerenkov effect with the Raman FEL; the device used an undulator for exciting the stimulated Raman effect in the presence of a dielectric medium that permits an additional frequency upshift.

Various theoretical studies were carried out in connection with the experiments mentioned above. Initially, theory was quantum-mechanical [125]; Sukhatme and Wolff [187] included the important feature of finite interaction length. However, Colson [46] showed how the two-wave FEL could be modeled as a dynamical problem with the pendulum equation; this model is particularly useful in understanding the nonuniform undulators and saturation effects. Sprangle et al. [182] were the first to calculate the growth rates of the weak pump Raman scattering process, and in a subsequent nonlinear analysis, growth rates and efficiencies were found for all the exponential processes [184]. The linear formulation of Kroll and McMullin [107] yielded the growth rates for the high-gain regime and also accounted for the low-gain two-wave scattering process. Further work on the high-gain system was reported by Hasegawa [96], including warm-beam effects, while Bernstein and Hirschfield [26] began a series of papers that treated the FEL as a boundary-value problem in traveling-wave amplification. Using a computational particle simulation code, Kwan et al. [112] analyzed the nonlinear problem, while Lin and Dawson [120] showed that high efficiencies—of order 25%—could be achieved by an appropriate profiling of ripple period and pump amplitude. These ideas were explored further, particularly in connection with the two-wave FEL, by Sprangle and Tang [185] and by Kroll, Morton, and Rosenbluth [6]. The latter theory is particularly valuable in establishing a connection between accelerator theory and FEL physics [139]. A highly useful contribution in interpreting short-pulse effects in the two-wave FEL using the concept of laser lethargy was made by AlAbawi et al. [16].

In the early 1980s, FEL activity had greatly increased; there were reports of particle trapping and efficiency enhancement experiments in FEL amplifiers [62, 177, 197]; amplification of visible light in a storage ring [55]; construction of far-infrared FELs at Bell Laboratories [167] and Frascati [53], using the recently improved microtron compact accelerator; and construction of storage-ring FEL facilities at many accelerator laboratories, among others, ACO in France, ADONE in Italy, and at Brookhaven and Stanford in the United States. Recently, workers at the French storage ring have reported an FEL oscillator at ≈ 6400 Å [29], and an FEL oscillator has been operated on the third harmonic of the FEL frequency near 0.5 μm [63] at Stanford; both these results depended on the use of nonuniform undulators. Energy-recovery experiments and preparations for a two-stage FEL were in progress in 1982 at Santa Barbara [67]. Encouraged by their success, plans were laid to construct FELs in the UV and soft X-ray spectral regions.

2

Conceptual Background

2.1 *Introduction*

The new physics in the FEL is the operation in the mode of stimulated scattering: this forms the primary distinction between earlier work (prior to 1976) and that which followed. The distinction is not merely a matter of the substantial increase in power which results when spontaneous emission is replaced by induced emission; we would be considering only a minor application of electrodynamics if this were the case.

The radiation of an isolated electron is given by a well-known single-particle formula (Section 2.4), and this radiation contributes to the ambient electromagnetic noise. This spontaneous radiation occurs independently of neighboring electrons and does not require an external, macroscopic EM field. Actual systems contain many electrons, and so it is not strictly true that the power output is a random superposition of the individually radiating electrons. Noise emitted by one group of electrons can be amplified by another group, under suitable conditions. Amplification of noise (superfluorescence, or loosely, superradiance) has been observed e.g., in high-gain gas lasers [83], and it can be observed in high-gain FELs as well. Given adequate feedback, the system becomes an oscillator, producing coherent radiation; this effect is well understood in quantum electronics (for a lucid introduction see A. E. Siegman's text). Our objective in this chapter is to identify the mechanism of gain or amplification.

We begin with simple models of electron motion in the undulator, and determine the spectrum of the scattered radiation in the single-particle limit.

The mechanism for gain is then identified in the hypothetical single-electron limit as well as in the collective limit when well-defined modes exist. The actual situation does not correspond to either example. The reader is therefore asked to read the caveats carefully before adopting these simplified models. Better models and quantitative analysis are to be found in Chapters 3, 4, and 6.

2.2 *Orbital Motion of an Electron in a Simple Undulator*

To begin, we take a simple approximation to determine the motion of an electron in the periodic magnetostatic field of an idealized undulator (Fig. 2.1). The field is represented as

$$B_\perp(z) = \sqrt{2}\, B_\perp \hat{y} \sin k_0 z \qquad (2.1)$$

where $k_0 = 2\pi/l_0$ and $B_\perp$ is the rms field amplitude. This field is unphysical, as it clearly cannot satisfy $\nabla \times B_\perp = 0$; we have omitted another field component in the axial (z) direction, and we have not accounted for the dependence of the field upon the transverse coordinates. This inconsistency is not serious as long as the electron remains close to the axis, i.e., the amplitude of its periodic motion is small compared with the scale length of the undulator field, k_0^{-1}. More accurate representations of the undulator fields and electron orbits will be given in Chapter 5.

The analysis starts with the electron equation of motion:

$$\frac{dp}{dt} = \frac{d}{dt}(\gamma m v) = \frac{e}{c}(v \times B) \qquad (2.2)$$

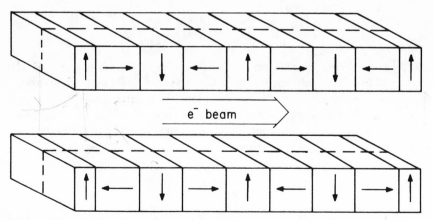

Figure 2.1 A linearly polarized undulator constructed of an assembly of samarium-cobalt permanent dipole magnets.

and since the magnetostatic field does no work on the electron, we set γ equal to a constant (this is only true if we neglect radiation) and simplify:

$$\gamma m \frac{d\boldsymbol{v}}{dt} = \frac{e}{c}(\boldsymbol{v} \times \boldsymbol{B}) \qquad (2.3)$$

which gives

$$\dot{v}_x = -\sqrt{2}\, \frac{v_z \Omega_\perp}{\gamma} \sin k_0 z$$

$$\dot{v}_z = \sqrt{2}\, \frac{v_x \Omega_\perp}{\gamma} \sin k_0 z \qquad (2.4)$$

We take $v_x \ll v_z$ so $z \approx v_z t \equiv v_{\parallel} t$, and define $\omega_0 = k_0 v_z$ and $\Omega_\perp = eB_\perp/mc$. The velocities are

$$v_x = \sqrt{2}\, \frac{v_{\parallel} \Omega_\perp}{\gamma \omega_0} \cos \omega_0 t$$

$$v_z = v_{\parallel} - 2 \frac{v_{\parallel} \Omega_\perp^2}{4\gamma^2 \omega_0^2} \cos 2\omega_0 t \qquad (2.5)$$

and the orbit is

$$x = \sqrt{2}\, \frac{v_{\parallel} \Omega_\perp}{\gamma \omega_0^2} \sin \omega_0 t$$

$$z = v_{\parallel} t - \frac{v_{\parallel} \Omega_\perp^2}{4\gamma^2 \omega_0^3} \sin 2\omega_0 t \qquad (2.6)$$

We have found the normalized rms amplitude of the "quiver velocity,"

$$\left(\frac{v_x}{v_z}\right)_{\text{rms}} = \frac{\Omega_\perp}{\gamma \omega_0} = \left(\frac{eB_\perp}{mc^2 k_0}\right)\frac{1}{\gamma} \equiv \frac{a_w}{\gamma} \qquad (2.7)$$

where a_w is the normalized rms vector potential of the undulator field. (It may be helpful to recall that $mc^2/e = 0.510$ MV $= 1700$ G-cm.) The orbital motion of the electron in this simple "linearly polarized" undulator is a sinusoidal curve having the same periodicity as the magnetic-field variation.

If an axial, constant magnetic field B_0 is imposed, the amplitude of the quiver motion becomes approximately

$$\left(\frac{v_x}{c}\right)_{\text{rms}} \approx \frac{\Omega_\perp/2}{|\Omega_0 - \gamma k_0 \beta c|} + \frac{\Omega_\perp/2}{|\Omega_0 + \gamma k_0 \beta c|} \qquad (2.8)$$

where $\Omega_0 = eB_0/mc$. The enhancement of the quiver motion for $\Omega_0 \approx \gamma k_0 \beta c$ is known as *magnetoresonance*.

The quiver motion of the electron is purchased at the expense of its forward motion, since the total energy is constant. Defining

$$\gamma = \left(1 - \frac{v_z^2}{c^2} - \frac{v_x^2}{c^2}\right)^{-1/2}$$

$$\gamma_{\parallel} = \left(1 - \frac{v_z^2}{c^2}\right)^{-1/2} \tag{2.9}$$

then

$$\overline{\gamma_{\parallel}^2} = \frac{\gamma^2}{1 + \gamma^2 \overline{v_x^2}/c^2} = \frac{\gamma^2}{1 + a_\omega^2} \tag{2.10}$$

where $\overline{v_x^2} = v_{x,\text{rms}}^2 = v_x^2/2$ for motion appropriate to Eq. (2.1). By using the rms amplitude for the electron quiver motion in this example of the "linearly polarized" undulator, one can use the same formulas for the case of the "circularly polarized" helical undulator, where the magnetic lines of force spiral around the axis with period l_0.

At this point it is useful to add a comment about the use of the word "undulator" as distinguished from the word "wiggler." It is now customary to refer to the magnetostatic, periodic device used in the FEL as the undulator; we have an undulator situation when the parameter a_ω used above in Eq. (2.10) is < 1. When $a_\omega \gg 1$, the same device is called a wiggler. We shall return to this point when we come to Eq. (2.21) in connection with the radiation.

2.3 The Relativistic Doppler Shift

Next, we wish to establish the wavelength dependence of the radiation emitted by an electron moving in the undulator [Eq. (1.1)]. Suppose we transform from the laboratory system of coordinates to a set attached to the moving electron (Fig. 2.2). The transformation is described by the theory of special relativity, and the appropriate relationships are the Lorentz transformations. In the laboratory frame, the magnetostatic periodic field has a certain amplitude $B_\perp$, period l_0, and wavenumber $k_0 = 2\pi/l_0$. In the electron frame (also called the "beam" or "rest" frame) the period is contracted: $l_0' = l_0/\gamma$ and also $k_0' = \gamma k_0$. The fields impinging on the electron in its own rest frame transform according to

$$B_\perp' = \gamma B_\perp \tag{2.11}$$

$$E_\perp' = \gamma \beta \times B_\perp \tag{2.12}$$

and since $\beta \approx 1$, $|B_\perp'| \approx |E_\perp'|$ and therefore these fields are very nearly

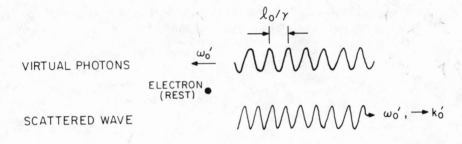

VIRTUAL PHOTONS

ELECTRON
(REST)

SCATTERED WAVE

REST FRAME

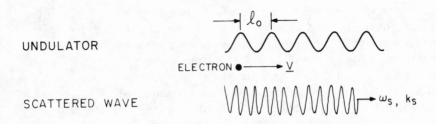

UNDULATOR

ELECTRON

SCATTERED WAVE

LABORATORY FRAME

Figure 2.2 Rest- and laboratory-frame quantities.

transverse. To a relativistic electron, the magnetostatic undulator field appears as an EM wave with high average intensity, and in this EM wave the electron quivers with the same frequency given by

$$\omega_0' = \gamma k_0 \beta_{\parallel} c \tag{2.13}$$

The electron is set into oscillation under the influence of the undulator field as described in Section 2.2, with appropriate transformation of the amplitude terms to obtain the rest-frame quantities. This transformation reveals an important piece of the underlying physics: an "inexpensive" magnetostatic field is transformed into an "expensive" high-intensity short-wavelength EM field. Interactions of the electron with this transformed field are necessarily electromagnetic, and waves resulting from this interaction remain electromagnetic, according to Einstein's postulate of relativity, when they are transformed back to the laboratory frame.

Since the undulator appears as an EM wave in the rest frame, we can imagine the electron Thomson-scattering this radiation, which can be thought of as a flux of "virtual" photons ("virtual" because there are obviously no photons associated with the magnetostatic undulator field in the laboratory frame). We can determine the frequency of the scattered radiation (ω_s) in the laboratory frame by Lorentz-transforming the frequency of the electron quiver motion (ω_0') as it appears in the rest frame:

$$\omega_0' = \gamma(\omega_s - \boldsymbol{k}_s \cdot \boldsymbol{v}) \qquad (2.14)$$

The radiation travels with speed c; hence $k_s = \omega_s/c$ and $k_0' = \omega_0'/c$. Investigating the direction where $\boldsymbol{k}$ is nearly parallel to $\boldsymbol{v}$ ($\approx v_\parallel$), with small angle θ between $\boldsymbol{k}$ and $\boldsymbol{v}_\parallel$,

$$\omega_0' = \gamma\omega_s(1 - \beta_\parallel \cos\theta) \qquad (2.15)$$

Since $\beta_\parallel \cos\theta \lesssim 1$ and $1 - \beta_\parallel \approx 1/2\gamma_\parallel^2$, then $\lambda_s = 2\pi c/\omega_s$ is

$$\lambda_s \approx l_0/2\gamma_\parallel^2 \qquad (2.16)$$

If we take into account the finite quiver motion, Eq. (2.10), as well as the possibility of off-axis propagation [factor $\beta_\parallel \cos\theta$], Eq. (2.15) becomes not Eq. (2.16), but more accurately

$$\lambda_s \approx \frac{l_0}{2\gamma^2}(1 + a_\omega^2 + \gamma^2\theta^2) \qquad (2.17)$$

It is apparant that the Doppler effect is quite useful in shortening the wavelength of the radiation: taking $l_0 \sim 1$ cm and using a modest energy beam, say 2 MV, λ_s falls in the submillimeter range; if the electron energy is of order 20–50 MV, then λ_s falls in the near IR. Considering on the other hand the case where radiation propagates antiparallel to the electron motion, the corresponding wavelength is $\approx 2l_0$, as one can verify by replacing the minus sign in eq. 2.15 with a plus sign.

The second point that one should bear in mind about the Thomson scattering process is that the scattering cross section is highly anisotropic. From quantum electrodynamics, neglecting terms of order $\hbar\omega/mc^2$, the invariant photon differential scattering cross section, when the incident beam of radiation is directed against the electron flow, is

$$\left(\frac{d\sigma}{d\Omega}\right)_{\text{photon}} \approx \frac{r_0^2}{\gamma^2}\frac{1}{[1 - \boldsymbol{\beta}\cdot\hat{\boldsymbol{n}}]^2} \qquad (2.18)$$

where $r_0 = e^2/mc^2$, and $\hat{\boldsymbol{n}}$ is the unit vector parallel to the scattered wave direction. For various backscattered angles, we get

$$\left(\frac{d\sigma}{d\Omega}\right)_{\text{photon}} \approx \begin{cases} 4\gamma^2 r_0^2 & (\theta = 0) \\ r_0^2/\gamma^2 & (\theta = 90°) \\ r_0^2/4\gamma^2 & (\theta = 180°) \end{cases} \qquad (2.19)$$

The total cross section does not depend on γ^2 (it is $\frac{8}{3}\pi r_0^2$), but it turns out the electron is a more efficient scatterer if we detect photons scattered in the same direction as the moving electrons.

Under circumstances where the undulator is a bona fide EM wave having frequency ω_i in the laboratory frame, we apply the Doppler formula twice, first to convert ω_i to ω_i' in the rest frame, and second to convert ω_s' into ω_s. We use $\omega_i' = \omega_s'$ in the rest frame—the same relationship used to calculate reflection from a "moving mirror" [203]—and obtain

$$\frac{\omega_s}{\omega_i} = \frac{4\gamma_\parallel^2}{1 + \gamma^2\theta^2} \qquad (2.20)$$

which is also the upshift in the scattered photon energy.

Radiation from the electron will occur in harmonics:

$$\lambda_{sp} = \lambda_s/p \qquad (2.21)$$

where p ($= 1, 2, 3, \ldots$) is the harmonic number. The spectrum of harmonics depends on the type of undulator. If $\theta = 0$, the electron emits only the fundamental harmonic if the undulator has a helical winding, whereas all the odd harmonics are present for a linearly polarized undulator, such as the one shown in Fig. 2.1. When the undulator is used as a wiggler, viz. $a_\omega \gg 1$, the spectrum of the system consists of broadband synchrotron radiation including many harmonics. This is useful for certain kinds of UV radiation sources, but it is not suitable for an FEL, where a coherent spectrum of the fundamental, either alone or possibly in conjunction with one or two harmonics, is desired.

The FEL equation (2.16) can be given a somewhat different interpretation if, instead of considering the frequency of the spontaneous radiation, one considers interaction of the electron with an external coherent beam of laser radiation: then l_0 and λ_s are determined, and the electron energy parameter γ is the variable. Let us denote the energy for which Eq. (2.16) applies by γ_r; it is the energy of the "resonance" electron and is characterized by v_r and β_r. In the electron rest frame, a ripple of the undulator field passes over the electron in a time

$$t' = l_0/\gamma v \qquad (2.22)$$

But in t' seconds, light will travel $ct' = l_0 c/\gamma v = l_0/\gamma\beta$ centimeters. We observe that in the electron frame the FEL wavelength is just

$$\lambda_s' = \frac{l_0}{\gamma\beta} \qquad (2.23)$$

Therefore, the electrons pass through one undulator period as one optical wavelength passes over them, providing β is such that γ satisfies Eq. (2.16)

—i.e., $\beta = \beta_r$ and $\gamma = \gamma_r$. We can write (2.16) in the original form

$$\omega_s = \frac{k_0 \beta_r c}{1 - \beta_r} = k_s c \qquad (2.24)$$

and then solve for β_r:

$$\beta_r = \frac{\omega_s}{\omega_s + k_0 c} = \frac{k_s}{k_s + k_0} \qquad (2.25)$$

Suppose we launch an electron along the axis of the undulator with the motion phased so that the relationships at $z = z_0$ are described by Fig. 2.3. Then, if the electron is resonant, that phase relationship is maintained throughout the device. As the electron moves down the axis, the phase of its motion advances by 2π when it has moved one undulator period. However, the wave phase must advance ahead of the electron, as shown, because not only is the velocity of light greater than the velocity of the electron, but the electron also moves on a curvy orbit along a path which is longer than that traveled by the radiation. The wave moves ahead of the electron by a phase proportional to

$$\frac{2\pi}{\lambda_s}\left[\frac{c-v}{v} + \frac{L_e - l_0}{L_e}\right] \qquad (2.26)$$

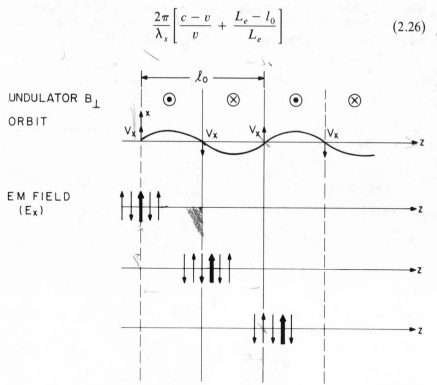

Figure 2.3 Relation between radiation and electron motion in resonance conditions; the undulator field is shown above. The boldface arrow in the field is a reference. (After Morton, [139]; © 1982 Addison-Wesley.)

One can calculate

$$L_e \equiv \int ds \tag{2.27}$$

along the electron orbit, using

$$ds^2 = dx^2 + dz^2 \tag{2.28}$$

and

$$\frac{v_x}{v_z} = \frac{a_\omega}{\gamma} \cos k_0 z \tag{2.29}$$

and find that Eq. (2.26) is

$$\frac{k_s}{2\gamma^2}(1 + a_\omega^2) \tag{2.30}$$

which will give the resonance relationship provided we set $\gamma = \gamma_r$ and equate (2.30) to the corresponding phase change $2\pi/l_0 = k_0$ experienced by the electron motion. However, if the electron is nonresonant, the phase slippage between the wave and the orbital motion is not 2π (as shown in Fig. 2.3), but is given instead by

$$\frac{d\psi}{dz} = k_0 - \frac{k_s}{2\gamma^2}(1 + a_\omega^2) \tag{2.31}$$

As $d\psi/dz = 0$ defines the resonance energy γ_r, for a different energy $\gamma = \gamma_r + \Delta\gamma$ Eq. (2.31) becomes

$$\frac{d\psi}{dz} = 2k_0\frac{\Delta\gamma}{\gamma} \tag{2.32}$$

Rewriting Eq. (2.31), we have

$$\frac{d\psi}{dz} = k_0 - \frac{k_s}{\beta} + k_s \tag{2.33}$$

so

$$\psi \approx (k_0 + k_s)z - \frac{k_s z}{\beta} + \text{constant} \tag{2.34}$$

but $z = \beta ct$ and $\omega_s = k_s c$ and therefore

$$\psi \approx (k_0 + k_s)z - \omega_s t + \phi \tag{2.35}$$

where ϕ is the laser wave phase. If ϕ is not constant, then it will appear as an additional term $d\phi/dz$ in Eq. (2.31).

2.4   *Spontaneous Radiation*

The power radiated by a single electron moving down a linear undulator can be calculated from the formulas of classical electrodynamics. Taking the geometry shown in Fig. 2.4, we find

$$\frac{dP_s}{d\Omega} = \frac{e^2}{4\pi c^3}\left(\frac{1}{\varkappa^2}\right)\left[a^2 - \frac{(1 - \beta^2)(\hat{n} \cdot a)^2}{\varkappa^2}\right] \qquad (2.36)$$

where $\varkappa = 1 - \hat{n} \cdot v/c = 1 - \beta\cos\theta$. Expanding Eq. (2.36) in the small angle θ,

$$\frac{dP_s}{d\Omega} \approx \frac{2e^2}{c^3}(\dot{v}_x)^2\frac{\gamma^6}{(1 + \gamma^2\theta^2)^3}\left[1 - \frac{4\gamma^2\theta^2}{(1 + \gamma^2\theta^2)^2}\cos^2\varphi\right] \qquad (2.37)$$

Despite the enhancement of power by the high exponent of γ, the amount of power radiated even by a very energetic beam of electrons is not large. Indeed, if the electrons were spaced uniformly along the beam, there would be no power emitted at all: when the beam has a steady current, for each electron emitting radiation with a given phase, there is another which radiates a field exactly out of phase, so that the two fields null each other. Incoherent radiation arises because the current is not exactly uniform—it is the fluctuations in particle current or density which account for net radiation, and it is well known that the mean-square size of these fluctuations is proportional to the average electron density. Hence the rms radiation field strength scales as $n^{1/2}$ rather than n, and the incoherent net power radiated scales as n.

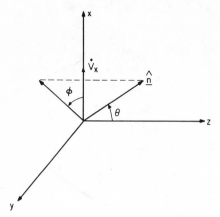

Figure 2.4   Coordinates of electron moving along z-axis for calculating radiation.

If the electrons spent an infinite amount of time in the undulator, the spectrum would be a delta function at wavelength given by Eq. (2.16). However, the electron experiences only N periods in the finite undulator length $L = Nl_0$. There is therefore a finite transit time $t = L/\beta_{\parallel}c$, and the power spectrum is broadened when one extracts the Fourier transform. (A similar situation, familiar to the reader, occurs when a beam of monochromatic light is reflected from a diffraction grating on which a finite number of grooves are illuminated.) Using the approach in Jackson's text [3], the power spectrum varies as (Fig. 2.5)

$$\frac{\sin^2(N\pi\xi)}{\xi^2} \tag{2.38}$$

where

$$\xi = \left[\frac{\omega}{k_0 c}(1 - \beta_{\parallel}\cos\theta) - 1\right] = \left(1 - \frac{\lambda_r}{\lambda}\right) = \left(1 - \frac{\gamma_r^2}{\gamma^2}\right) \tag{2.39}$$

in which λ_r is the resonant wavelength for a given energy γ, and γ_r is the resonant energy for a given wavelength λ. This is the spectrum of the spontaneous or noise radiation, and it has fractional width $\sim 1/N$.

Detailed radiation spectra for both the linear and helical undulator cases, including off-axis and harmonic radiation, are quoted in [48]. For equal undulator fields ($B_\perp$), the helical-wound undulator will radiate twice as much power as the linearly polarized undulator.

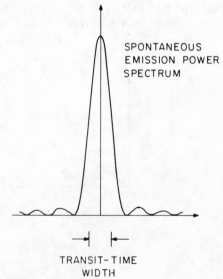

SPONTANEOUS
EMISSION POWER
SPECTRUM

TRANSIT–TIME
WIDTH

Figure 2.5 Spectrum of spontaneous power; the abscissa is proportional to ξ.

2.5 Stimulated Scattering

Inspecting Fig. 2.3, one can understand that by adjusting the phase ψ between the electron orbital motion and the EM wave, one can arrange that the electron will exchange energy with the field. The rate at which the electron's energy changes is

$$mc^2\frac{d\gamma}{dt} = mc^2 v_z \frac{d\gamma}{dz} \qquad (2.40)$$

which can be equated to the power input from the fields. The latter occurs via the interaction of the axial electron velocity with a synchronous axial force. The latter is the *ponderomotive* force, and it is responsible for changing the axial electron position and velocity. It is

$$-|e|\left(\frac{v_\perp}{c} \times B_s\right) \qquad (2.41)$$

where B_s ($\approx E_s$) is the magnetic component of the EM wave. (The synchronism can be appreciated by consulting Fig. 2.3.) Thus, using Eq. (2.5), the rate of increase of average energy is

$$mc^2\frac{d\gamma}{dz} = -eB_s\left\{\left(\frac{eB_\perp}{mc^2 k_0}\right)\frac{1}{\gamma}\right\}\sin\psi \qquad (2.42)$$

We define

$$a_s = \frac{eB_z}{k_s mc^2} = \frac{eE_s}{k_s mc^2} \qquad (2.43)$$

and therefore

$$\frac{d\gamma}{dz} = -\left(\frac{k_s a_s a_\omega}{\gamma}\right)\sin\psi \qquad (2.44)$$

From eq. 2.44 it is apparent how the undulator couples the electron's motion to the EM wave, and in this way the electron can interact with a short-wavelength optical wave. Depending on the choice of γ and ψ, an electron injected into the undulator can be accelerated or decelerated by the ponderomotive force. Given a suitable interval, this may lead to bunching of the electrons in the beam. For an undulator in which the period and amplitude are constant, an electron injected at the resonance energy will exit with the same energy, and so $\psi_r = 0$; but for $\Delta\gamma = \gamma - \gamma_r$, there will be loss or gain of energy to the optical wave. To cover all eventualities, we can define a resonant phase ψ_r associated with the resonant energy γ_r, in which case eq. 2.44 becomes $d(\Delta\gamma)/dz = -k_s a_s a_\omega/\gamma_r[\sin\psi - \sin\psi_r]$. We shall examine this in more detail in Chapter 3.

Let us now relate this physics to stimulated emission or absorption. Suppose an electron, moving along an undulator, interacts with a co-propagating external EM wave—for example, a laser beam. Choose the energy of the electron and the wavelength of the laser so that the latter corresponds to the wavelength of the spontaneous radiation. The process by which a relativistic electron interacts with the "virtual" photons of the undulator is essentially the same as its interaction with an electromagnetic pump, provided the latter wavelength is $2l_0$ [compare Eqs. (2.17) and (2.20)], and so the undulator and the laser waves can be treated on the same footing. It is plausible that the light wave might be amplified at the expense of the electron energy.

In the electron rest frame the undulator and laser waves impinge upon the electron from opposite directions (Fig. 2.6). There are two possibilities: (I) the electron scatters the undulator virtual photons ahead and loses momentum, or (II) the electron scatters the laser photon backward and gains momentum. The Compton diagram (Fig. 2.6c) can be used to compute the frequency shift $\Delta\omega'$ between the incident (ω_i') and scattered (ω_s') waves in, say, the example of case (II) [set $\phi_c' = 180°, \theta_c' = 0°$]. Using conservation of momentum and energy, and eliminating the electron momentum, one obtains the frequency shift $\Delta\omega' = \omega_i' - \omega_s'$ due to the electron recoil:

$$\frac{\Delta\omega'}{\omega'} = \frac{2\hbar\omega'}{mc^2} \qquad (2.45)$$

However, the situation in case I is just the opposite to that of case II (the arrows can be reversed in the Compton diagram). Equation (2.45) therefore shows that these two processes occur at different wavelengths. Case I pertains to light amplification and electron deceleration, while case II pertains to light absorption and electron acceleration. Stimulated emission therefore does not occur at the same frequency as stimulated absorption.

The fractional frequency shift (2.45) is $2\hbar\omega/\gamma mc^2$ in the laboratory frame, and this is very small compared with the width of the spontaneous-emission line. If $\Delta\omega'$ were zero, the probability of induced emission would exactly equal the induced absorption. The net gain can be calculated by subtracting two spontaneous emission curves, slightly displaced, as shown in Fig. 2.6d. The striking result is that, unlike a conventional laser, the gain function of an FEL is asymmetrical with respect to the center "resonance" frequency (in the atomic laser the gain—due to level overpopulation—resembles the spontaneous-emission spectrum, and is symmetrical about the resonance frequency). Positive gain in Fig. 2.6d corresponds to stimulated emission, while negative gain corresponds to stimulated absorption, and the asymmetry occurs with reference to the resonance wavelength in Eq. (2.25).

The argument given above might lead the reader to conclude that FEL gain is quantum-mechanical in nature, since electron recoil seems to be

LABORATORY FRAME   REST FRAME

laser beam (L)   laser (L′)   undulator (U′)
                                        (virtual photons)

e → v_{11}, γ_{11}

(Ⅱ)   e   (I)

undulator (U)

(a) INITIAL STATE

recoil   U′   U′
v′ e   forward scatter   backscatter   recoil
                                        e   v′
(I)   L′   L′   (Ⅱ)

(b) FINAL STATE-REST FRAME

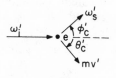

(c) COMPTON DIAGRAM

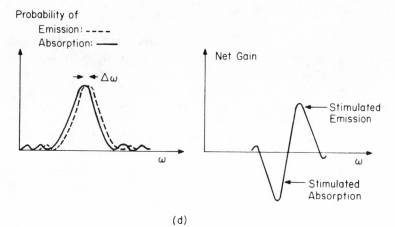

Probability of
Emission: ----
Absorption: ——

Δω

ω

Net Gain

Stimulated
Emission

ω

Stimulated
Absorption

(d)

Figure 2.6 Compton scattering interaction for a single electron: FEL laser and accelerator cases. The scattering diagram is drawn for case (II).

necessary to give finite gain. In the limit $\hbar \to 0$, the electron recoil vanishes together with the quantum-mechanical single-particle FEL gain. But, with *finite* undulator length, we have not yet allowed for charge bunching, which emerges from consideration of the motion of an ensemble of electrons (Chapter 3). A classical prediction for FEL gain, resembling Fig. 2.6d, is recovered in the theory for an undulator of finite length. Although we shall omit quantum-mechanical considerations in this book, many of the early theoretical FEL papers used quantum treatments. Later work [19] used a single-particle Hamiltonian and a quantum treatment which yielded the classical result of Colson [48] in the small-signal limit. The basic feature, that the gain spectrum is the derivative of the spontaneous-emission spectrum, remains true for two-wave FELs in the limit of small-signal-gain classical theory, where the EM fields can be approximated by plane waves [49]. This result, which will be encountered again in Chapter 3, is frequently referred to as Madey's theorem [127]; it has been verified quantitatively in a very carefully done experiment mentioned in Chapter 8 [55].

Given net positive gain, the analysis of an FEL proceeds like that of a traveling-wave amplifier. Electron bunching will progress down the undulator; the scattered virtual photons of the undulator appear in the laboratory frame as a spatially growing, Doppler-shortened wave. As the EM wave grows, the bunching is intensified (via the ponderomotive force), and the process bootstraps. The energy invested in the growing EM field is obtained from the electron streaming energy, at least in the limit where the ponderomotive potential is small and the space-charge field energy is weak. More detailed calculations of the FEL gain will be made in the next two chapters.

2.6 Effect of Electron Space Charge

So far our discussion of the FEL has been concerned with the interaction between the electron and the scattered wave. However, as the axial bunching of the electron space charge by the ponderomotive force develops, at some point the accumulation of space charge must become important, especially for high-current-density electron beams. Under certain circumstances it is possible to ignore the electron space charge (it turns out this simplifies some of the physics), yet one cannot conclude that the space-charge effects are merely annoying complications. On the contrary, they offer an opportunity to improve the FEL gain or efficiency under proper conditions. In this section we shall identify some of the qualitative physics which becomes important when the electron-beam space charge must be accounted for.

In the electron rest frame we consider the simple uniform state of a cold electron gas. Owing to the repulsive forces within this medium, an equilibrium must be provided by a collection of external forces. Even in tenuous beams, provision must be made (guiding magnetic field, electrostatic or magnetostatic focusing devices, etc.) so that the electron beam may be transported through the FEL while maintaining its integrity as a cold, relativistic stream of charges. Let us suppose a stable equilibrium has been provided. This implies that if we perturb the equilibrium, a restoring force is produced, and that only stable oscillatory motion of the electron fluid is possible.

Examples of perturbations of the beam equilibrium are easy to specify. We might ripple the surface boundary of the beam, or—more relevant to the FEL—we might begin by bunching the space charge periodically. How does the system respond? The energy in the perturbation flows out into the characteristic modes of motion of the system, and appears as waves.

Waves (collective motion) in an electron gas in zero magnetic field have two classifications: an EM mode ($\nabla \cdot E = 0$) and an electrostatic (ES) mode ($\nabla \times E = 0$). The former permits light to move through the electrons under certain conditions which are given by the *dispersion relation*

$$\omega^2 = k^2 c^2 + \omega_p^2 \tag{2.46}$$

where $\omega_p^2 = 4\pi n e^2/\gamma m$ defines the invariant plasma frequency for a relativistic system. This relates frequency and wavelength or wavenumber and is used to define the refractive index

$$\mu = \left(1 - \frac{\omega_p^2}{\omega^2}\right)^{1/2} = kc \tag{2.47}$$

which reveals that (a) the phase velocity of EM waves is slightly larger than c, since $\mu < 1$, and (b) the light-wave frequency must satisfy $\omega > \omega_p$. When a uniform magnetic field is imposed, light may propagate along its direction in two modes, the right- or left-hand circularly polarized waves, which have refractive indices

$$\mu_R = \left(1 - \frac{\omega_p^2}{\omega(\omega - \Omega_0)}\right)^{1/2} \tag{2.48}$$

$$\mu_L = \left(1 - \frac{\omega_p^2}{\omega(\omega + \Omega_0)}\right)^{1/2} \tag{2.49}$$

The case $\omega = \Omega_0$ for the right-hand wave is known as "cyclotron resonance."

The ES solution is also very simple: for a cold uniform electron gas,

$$\omega = \omega_p \tag{2.50}$$

and this applies also to the case of the magnetized medium, provided the

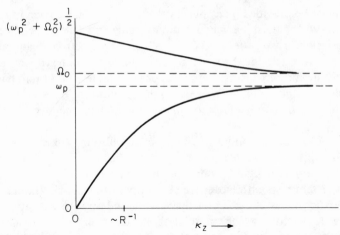

Figure 2.7 Dispersion diagram for cyclotron wave (above) and space-charge wave (below); effect of radial geometry is shown at small k.

electron oscillation is parallel to the lines of force. Should the beam have finite radial structure, the case is more complicated, because the electrostatic fields extend radially as well as axially and this drives motion transverse to the magnetic field. Two modes can then occur: the space-charge mode and the cyclotron mode (Fig. 2.7). For the present we shall assume the lateral extent of the beam is large, and we shall not discuss the cyclotron waves further until Chapter 6. If the electron gas is warm, the plasma oscillations become dispersive:

$$\omega^2 = \omega_p^2 + 3k^2 v_T^2 \tag{2.51}$$

The phase velocity slows and approaches the thermal velocity $v_T^2 = T/m$ of the electrons as k becomes large. A bound on the collective behavior is set when k approaches k_{Debye}, or $\lambda \approx \lambda_D$, where

$$\lambda_D \equiv v_T/\omega_p \tag{2.52}$$

In that case the wave motion is dissipated by a collisionless mechanism known as Landau damping.

When a cold electron gas moves with speed v (as in the beam example), a Doppler shift is introduced between waves as observed in the rest frame and as observed in the laboratory frame. Transforming Eq. (2.50) with the aid of Eq. (2.14),

$$\omega = k_z v \pm \frac{\omega_p}{\gamma} \tag{2.53}$$

for the axisymmetric beam space-charge modes. The phase velocity of these modes is

$$v_\phi = \frac{\omega}{k_z} = v \left(\frac{\omega}{\omega \mp \omega_p/\gamma} \right) \tag{2.54}$$

One wave is moving faster than the electrons while the other is moving slower. These two modes, the fast and slow space-charge waves, are shown in Fig. 2.8a, with the light line superimposed. An intersection of the fast space-charge wave with the light line is possible: this means the fast

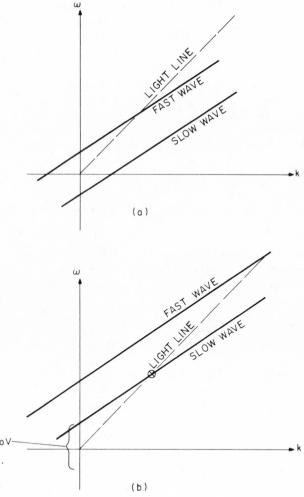

Figure 2.8 Effect of undulator on space-charge waves: above, no undulator; below, with undulator.

space-charge wave and an EM wave can move synchronously. However, the fast wave cannot extract energy from the electrons. This would be possible if the electron could gain energy so that it could move at the same speed as the wave phase velocity; but since the electron has no way to acquire this extra energy, it is not possible to feed the kinetic energy of the electron into a growing fast space-charge wave. The situation is different with regard to the slow space-charge wave: by various processes an electron might dissipate some of its kinetic energy, and in slowing down, fall into synchronism with the slow space-charge wave. As the energy of the electron falls, it interacts with the wave in such a way as to feed some of its kinetic energy into the wave fields. The amplitude of the slow space-charge wave will grow. The fast space-charge wave is a stable perturbation of the beam, whereas the slow space-charge wave is an unstable perturbation. Nevertheless, although we may have a slow space-charge wave growing, no radiation results, because the slow wave does not intersect the light line: there is no synchronous interaction (same ω and k_z) for the EM and the slow ES wave.

In order to couple the EM and ES waves, an undulator is used (there are other methods available as well). In the rest frame of the electron, the undulator frequency $k_0'v$ changes Eq. (2.50) to

$$\omega' = k_0'v \pm \omega_p \tag{2.55}$$

and, applying the Doppler formula (2.14), we get

$$\omega = (k_z + k_0)v - \frac{\omega_p}{\gamma} \tag{2.56}$$

for the slow space-charge wave as modified by the undulator. From Fig. 2.8b, it is apparent that this modified wave can now intersect the light line. The intersection is obtained by using a simplified dispersion relation for light, $\omega_s = k_s c$:

$$\omega_s = 2\gamma_{\parallel}^2 \left(k_0 \beta c - \frac{\omega_p}{\gamma} \right) = k_s c \tag{2.57}$$

where again we have used $2\gamma_{\parallel}^2 \approx (1 - \beta_{\parallel})^{-1}$; this "modified" FEL formula applies to what is now understood to be the *three-wave* FEL. The unstable slow space-charge wave drains energy from the electron stream by an energetically favored instability, and then the slow space-charge wave itself can interact with EM radiation at the frequency defined by the FEL dispersion relation. It is evident that the physics involved is rather different from the two-wave FEL, but one is led to a similar result.

The three-wave approach can be understood best by stepping into the electron rest frame and considering the interaction of (1) the undulator (EM) wave (ω_0', k_0'), (2) the scattered (EM) wave (ω_s', k_s'), and (3) a beat, or "idler," wave. The idler arises from the ponderomotive bunching of the electrons through the interaction of the electron quiver velocity—induced by the undulator—and the scattered wave magnetic component B_s'. The idler is an ES wave, and it is also apparent by inspecting the nonlinear ponderomotive force term

$$- \frac{|e|}{c} \left[v_\perp' \left(\omega_0' \right) \times B_s' \left(\omega_s' \right) \right] \tag{2.58}$$

that it contains Fourier components at frequency $\omega_0' \pm \omega_s'$. The unstable interaction occurs at $\omega_0' - \omega_s'$, so we have

$$\omega_i' = \omega_0' - \omega_s'$$

$$k_i' = k_s' + k_0' \tag{2.59}$$

for the idler frequency and wavenumber (where the wavenumber relation obtains in the backscattering geometry). The idler or "beat" wave will always occur, but we expect that the interaction will be enhanced whenever ω_i' corresponds to a "natural" resonant frequency of the system, viz., the plasma wave, $\omega_i' = \omega_p$. Since $|k_s'| \approx |k_0'|$, it follows that $|k_i'| \approx 2k_0'$. The three-wave interaction, $\omega_p = \omega_0' - \omega_s'$, is referred to as "stimulated Raman scattering" (Fig. 2.9). The parallelogram relating the three waves follows from Eq. (2.59). The electron recoil, which we found was necessary for FEL gain in Section 2.5, occurs in the three-wave interaction in connection with the release of a "plasmon" to the space-charge wave; thus gain occurs with an exponential growth coefficient not depending on the undulator length (Chapter 4).

The model we have just presented is an oversimplified one involving the uncoupled modes of the system. The actual situation is more subtle; as a

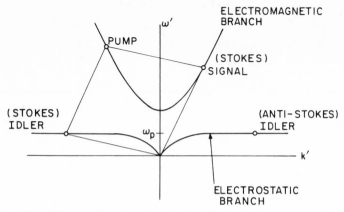

Figure 2.9   Three-wave scattering relationship in the electron rest frame.

coupling of modes (i.e. intersections in Fig. 2.8b) occurs, the different modes lose their clear EM or ES character. It is found, for example that the idler wave becomes largely transverse, and hence the "ponderomotive" force term [Eq. (2.58)] is not the sole cause of the interaction. Bunching in phase is the more appropriate description, whereas axial bunching is less important.

The two- and three-wave processes occur simultaneously in the small-signal limit, and competitively in the large-signal limit. We shall gain some perspective on this in Chapter 6; for now we point out that the three-wave process is a convectively growing instability and causes a signal present at the input of the FEL to exponentiate as it moves along the undulator, whereas the two-wave mechanism does not. Three-wave FELs usually have high gain. To decide whether the effects of space-charge waves are important or not, we imagine that the electron-beam system senses the collective behavior of the medium by trying to establish a plasma oscillation. We can count the plasma oscillations in the rest frame as the undulator passes by, and if the number is large, collective effects will be important:

$$\omega_p t_0 = \omega_p L/\gamma c \gg 1 \tag{2.60}$$

since the characteristic time in the electron rest frame is $L/\gamma\beta c$. The scaling relationship, $n \propto \gamma^3$ establishes the following requirement: collective effects in the FEL are important only at high electron density and low beam energy. However, because the two-wave FEL gain is low, collective effects may be of marginal use even when the beam energy is high. This will be studied analytically in Chapter 4.

3

FEL Theory, Part I:
The Two-Wave FEL

3.1 *Introduction*

Having surveyed the conceptual basis of the FEL, we now seek a more detailed and quantitative understanding. For the experienced researcher, this can be obtained by writing down a set of equations appropriate to the physical situation; these equations usually require numerical processing or certain simplifications which restrict their range of validity. Indeed, we shall adopt such an approach later in Chapter 4 and in Chapter 6; however, in this chapter we are dealing with classical, single-particle phenomena, and intuition leads one to expect that a simple physical model should lead to some clear, approximate predictions and good insight into FEL physics. In other words, we shall approach the complexities of the FEL with the hope of uncovering something which is at least as comprehensible as, say, the Bohr model of the atom is to the student of atomic physics.

A brief list of the simplifications we shall adopt is as follows:

1. The electron beam shall be cold; in particular, it has no intrinsic spread of parallel velocities.
2. The dimensions of the beam are infinite, but the undulator has finite length.
3. The FEL will operate in the amplifier mode, that is, we assume there is a strong, monochromatic EM wave already present, which will

interact with the electrons. We are not literally starting from "noise level."

4. The electrons are distributed uniformly along the beam before it enters the undulator, and there are no important effects due to statistical irregularities or spontaneous radiation, until Section 3.5.

5. The orbits of the electrons are as given in Chapter 2.

6. The various waves are written in the plane-wave approximation, rather than as resonator modes or Gaussian waves.

7. Quantum-mechanical effects are not considered further.

8. The space-charge fields developed within the beam as a result of the FEL interaction will be neglected.

The models used in this chapter were developed by Colson [46] and Morton [139] and have been selected for presentation here because they involve models familiar to most physicists. The more sophisticated approach to the problem of laser lethargy and startup is drawn from the work of Sprangle et al. [186].

3.2 Linear-Accelerator Model

A very helpful observation was made by Morton [39] in connection with FEL theory—namely, under certain conditions, there was an analogy between the physics of the FEL and the physics of rf linacs and storage rings. The analogy permits the transfer of useful accelerator theory and practice to the new subject of the FEL. As accelerator physics forms a common intellectual base for most physicists, we draw upon this in our first exposure to FEL theory. We shall begin by considering the interaction of an rf accelerating field with a relativistic electron, referring to the schematic in Fig. 3.1.

In Fig. 3.1a is shown a section of a linear accelerator: the electron is accelerated by the axial component of electric field as it passes through a resonant cavity. The spacing between cavities is chosen so that the rf field advances by one cycle as the electron moves between adjacent cavities. In the example of the ring, Fig. 3.1b, one cavity will do; once the electron's energy is at a fixed value, the rf energy serves to replenish the losses due to synchrotron emission. The diameter of the ring depends on the magnetic field guiding elements and the electron momentum. For the equilibrium orbits a given energy γ will correspond to a given circumference l, while another energy $\gamma + \Delta\gamma$ will correspond to $l + \Delta l$. (The reader who wants a good review of storage-ring physics should read the excellent chapter by Sands [13].) Perturbation from the equilibrium orbit causes a periodic motion, which is stable for a certain range of externally controlled parameters. If either the particle velocity or its orbital length changes, then the orbital frequency ω_0 must also change: the relation is

$$\frac{\Delta\omega}{\omega_0} = \frac{\Delta v}{v} - \frac{\Delta l}{l} = \frac{\Delta\gamma}{\gamma}\left(\frac{1}{\gamma^2} - \alpha\right)\frac{1}{\beta^2} \tag{3.1}$$

where α is known as the momentum compaction factor [13]. If the rf field in the cavity is oscillating at frequency $\omega_0 = \omega_{rf}$, the incremental change in the phase ψ of the rf field between successive passages of the electron through the cavity, per unit circumferential length, is

$$\frac{\Delta\psi\,(\text{per revolution})}{2\pi R} \equiv \frac{d\psi}{dz} = -\frac{1}{\beta^2 R}\left(\frac{1}{\gamma^2} - \alpha\right)\frac{\Delta\gamma}{\gamma} \tag{3.2}$$

where z is measured along the equilibrium orbit.

Acceleration or deceleration of the electron in the cavity depends on the phase angle between the electron velocity and the component of electric field parallel to the motion. Since

$$nmc^2\frac{d\gamma}{dt} = \boldsymbol{j} \cdot \boldsymbol{E} \tag{3.3}$$

then

$$mc^2\left(\frac{d\gamma}{dz}\right)v_z = ev_z\overline{E}_z \sin\psi \tag{3.4}$$

or

$$\frac{d\gamma}{dz} = \frac{e\overline{E}_z}{mc^2}\sin\psi \tag{3.5}$$

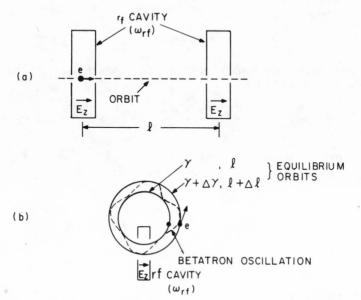

Figure 3.1 Orbits in an rf accelerator: linac section (a) and storage ring (b).

Turning our attention to the ring, electrons will radiate as they move around the orbit, and the electron must gain the appropriate amount of energy from the rf field to remain synchronous. This will occur when ψ is adjusted to a certain value, ψ_r, so that $d\gamma_r/dz$ just balances the losses:

$$\frac{d\gamma_r}{dz} = \frac{e\overline{E}_z}{mc^2} \sin \psi_r \tag{3.6}$$

If ψ is not at ψ_r, the electron will gain or lose energy. Then

$$\frac{d}{dz}(\Delta\gamma) = \frac{e\overline{E}_z}{mc^2}[\sin \psi - \sin \psi_r] \tag{3.7}$$

Equation (3.7) for $d(\Delta\gamma)/dz$ and Eq. (3.2) for $d\psi/dz$ are the standard equations used in accelerator design.

3.3 *Wave-Electron Interaction in the FEL*

The Pendulum Equation

In Section 2.3 the resonant electron energy was defined in terms of the phase between the optical electric field and the transverse component of electron velocity:

$$\frac{d\psi}{dz} = 2k_0 \frac{\Delta\gamma}{\gamma}$$

$$\Delta\gamma = \gamma - \gamma_r \tag{3.8}$$

Associated with the resonant energy γ_r is a resonant phase angle ψ_r, obtained from Eq. (2.44):

$$\frac{d\gamma_r}{dz} = -\frac{k_s a_s a_\omega}{\gamma_r} \sin \psi_r \tag{3.9}$$

An electron will gain or lose energy γ if it is injected with $\psi \neq \psi_r$. The analogy between the accelerator and the FEL is made by the substitutions

$$\frac{e\overline{E}_z}{mc^2} \rightarrow -\frac{k_s a_s a_\omega}{\gamma_r}$$

$$-\frac{1}{\beta^2 R}\left(\frac{1}{\gamma^2} - \alpha\right) \rightarrow 2k_0 \tag{3.10}$$

The distinction between FEL and accelerator physics is that in the latter the electric field is parallel to the primary electron motion; in the FEL the undulator causes a transverse component of motion which permits coupling with the optical field.

Referring to Eqs. (3.7) and (3.9), the electron energy in the FEL will vary about γ_r if ψ is changed from ψ_r:

$$\frac{d}{dz}(\Delta\gamma) = \frac{-k_s a_s a_\omega}{\gamma_r}[\sin\psi - \sin\psi_r] \qquad (3.11)$$

We now combine the two basic equations (for $\Delta\gamma$ and ψ) into one equation:

$$\frac{d^2\psi}{dz^2} = 2k_0 \frac{d}{dz}\left(\frac{\Delta\gamma}{\gamma}\right) = 2k_0\left(\frac{-k_s a_\omega}{\gamma_r^2}\right)[\sin\psi - \sin\psi_r] \qquad (3.12)$$

or

$$\gamma \cong \gamma_r$$

$$\frac{d^2\psi}{dz^2} = -\Omega_L^2[\sin\psi - \sin\psi_r] \qquad (3.13)$$

which is known as the *pendulum equation*.

A first integral of the pendulum equation will give $d\psi/dz$ in analytic form, whereas a second integration involves the elliptic integral functions. The characteristic period (in centimeters) of small-amplitude vibrations for $\psi_r = 0$ is

$$\frac{2\pi}{\Omega_L} = \frac{l_0}{2\sqrt{a_s a_\omega}} \equiv L_B \qquad (3.14)$$

Here L_B is the "bounce" distance, and c/L_B is the "bounce" frequency. The analog of the bounce frequency in Quantum Electronics is the Rabi frequency. The electron can be imagined to move under the influence of a spatially periodic ponderomotive potential which depends on $a_s a_\omega$. [The dynamical problem is also derivable from a Hamiltonian $= k_0(\Delta\gamma)^2/\gamma_r + F(\psi)$ where $F(\psi) = -k_s a_s a_\omega(\cos\psi + \psi\sin\psi_r)/\gamma_r$.] The ponderomotive wave moves slightly slower than the electrons, with phase velocity $\omega/(k_s + k_0)$, so electrons will give energy to the wave as they fall into synchronism. If the electron is trapped in a ponderomotive potential well, it will execute nearly harmonic motion at the bounce frequency. The motion of the electrons trapped in a potential trough of the wave will appear as a set of sidebands on the Fourier spectrum. Energy present in the wave field interacts with the electrons and can appear in these sidebands. In FEL research the periodic motion in the ponderomotive wave is also referred to as the "synchrotron oscillation".

For what follows in this section, we shall take $\psi_r = 0$ [the term $\sin\psi_r$ represents a constant torque on the "pendulum"]; this defines an FEL where the undulator amplitude and period are fixed, as well as γ_r. Then eq

3.13 describes the motion of a pendulum in which the pendulum angle is of arbitrary size; for small angle ψ, the motion is of course simple-harmonic. We can understand some important FEL physics by examining some well-understood features of the pendulum equation. In Fig 3.2, the pendulum phase diagram, the coordinate $\psi' = d\psi/dz$ is plotted versus ψ. Above the phase diagram is a representation of the ponderomotive wave. It is useful to note, from Eq. (3.8), that $d\psi/dz$ is proportional to $\Delta\gamma$ (the energy change), and also—from Eq. (2.35)—that ψ is related to z or t in connection with the electron's motion. Therefore, although we keep in mind the motion of the pendulum, we are actually following the motion of a particular electron characterized by a particular energy (or ψ') and location (ψ) at any point along the undulator.

In Fig. 3.2, the width of the "closed" region in the phase space depends on Ω_L, i.e., it depends on the amplitude of the undulator field and the EM field: large fields will expand the closed region in (ψ', ψ) space. The points ($p\pi$, 0) are "critical" points. Motion near p an even integer corresponds to stable, simple harmonic motion at the bounce frequency. Motion about the points corresponding to p an odd integer is unstable—here the "pendulum" is at the top of its arc. Motion outside the separatrix would describe the pendulum swinging completely around.

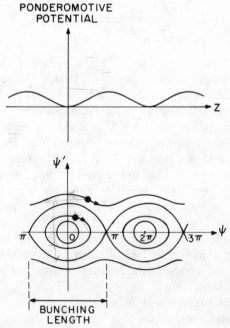

PONDEROMOTIVE
POTENTIAL

Figure 3.2 FEL phase space (below) and ponderomotive wave potential (above).

This type of FEL [we call it the "two-wave" or "single-particle" or "Compton" FEL] can operate in two different ways. If the amplitude a_s is large enough—i.e., the scattered or amplified wave is intense—a substantial fraction of the electrons in the beam may become trapped in the potential well of the ponderomotive wave. The gain is due to the extraction of kinetic energy as the electrons slosh in the potential troughs of this wave. The maximum gain will occur when the electrons are injected near the "top" of the closed region in the phase diagram (Fig. 3.3, discussed in the next subsection) and are extracted near the "bottom." This requires half a cycle of bounce motion, corresponding to cutting the undulator length at one-half a bounce length. We shall discuss this example first below.

On the other hand, an FEL may operate at low power, or it may have to start at nearly zero amplitude. Then the ponderomotive wave is too small to trap the electrons. The closed area in the phase space of Fig. 3.2 is very small, and most electrons are on open orbits. The gain is due to the perturbation of the beam electrons by the ponderomotive wave (Fig. 3.5, discussed in the subsection after next [47]), which results in electron bunching. The interchange of energy between the wave and the electrons is oscillatory with period equal to that of the wave. The difference between these two conditions actually is made by selection of different orderings of the various field amplitudes.

Large Amplitude—Approximate Result

In order to see how one can extract energy from the electron stream in an FEL, refer to Fig. 3.3 [139]. In this computation, based on the pendulum equation, cold electrons (dotted line, Fig. 3.3a) were injected with energy near the top of the phase-space separatrix (γ_{inj}), above the resonant energy (γ_r). Owing to the injection of electrons with different phases, some will be accelerated and some will be decelerated [Eq. (3.9)], so that after a short time they will all have different energies. However, as the electrons are trapped and begin to slosh in the ponderomotive "bucket," the average energy $\bar{\gamma}$ of the ensemble changes (Fig. 3.3b, c). The maximum gain— obtained when the interaction is stopped at about $\frac{1}{2}$ bounce period— corresponds to a minimum of the ensemble average energy (Fig. 3.3b). The difference between the initial and final average energy of the ensemble ($\gamma_{inj} - \bar{\gamma}$) is given to the EM scattered field. If γ_{inj} is too far from γ_r, electrons will not be trapped, and the interaction will be comparatively weak. If we wait a full bounce period (Fig. 3.3c), the average energy of the ensemble approaches its initial energy, $\bar{\gamma} \approx \gamma_{inj}$. On the other hand, the bouncing is periodic, so we can recover net gain again by waiting another half bounce period, and so on. If one chooses $\gamma_{inj} < \gamma_r$, in $\frac{1}{2}$ bounce period the ensemble will be entrained and elevated in the bucket so that $\bar{\gamma} > \gamma_{inj}$—energy has then been extracted from the ponderomotive wave (and therefore from the scattered wave). By varying the injected energy and keeping account of the net transfer of electron energy to the wave, we trace out a gain diagram, sketched in Fig. 3.4, which resembles Fig. 2.6d. The energy loss of the electron, $\Delta\gamma = \gamma_{inj} - \bar{\gamma}$, is the gain of the EM wave; the abscissa, $\gamma_{inj} - \gamma_r$, is proportional to a frequency difference defined by the two parameters of the system—the undulator period and the period of the scattered wave (we call this the "beat" frequency in Section 2.6). For a certain value of $\gamma_{inj} - \gamma_r$, or equivalent beam frequency, we obtain optimum gain according to the physical principle outlined above. Another important point to keep in mind is that the process of electron injection followed by gain is accompanied by a spread in electron energies.

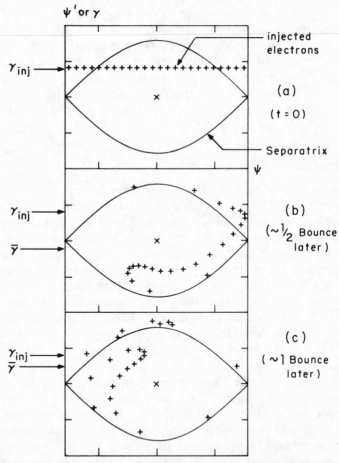

Figure 3.3 Injection of electrons into a ponderomotive bucket [139]: a, b, c, show situation at $t = 0$ and subsequent intervals. Dots represent the electrons. © 1982 Addison-Wesley.

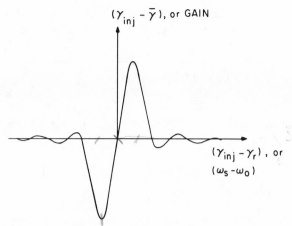

Figure 3.4 FEL gain function.

We now *estimate* the energy extracted from the electron stream under optimum conditions, that is, when the interaction is allowed to go on one-half bounce period or one undulator length. The energy loss is expressed as a change in γ, $\Delta\gamma_L = \gamma_{inj} - \bar{\gamma}$, and is conveniently written as the fraction $\Delta\gamma_L/\gamma$. The idea is to evaluate the energy, from Eq. (3.11), at roughly one-half bounce period. Then the argument of $\sin\psi$ has changed so that $\Delta(\sin\psi) \approx 2$. Defining the bounce distance as $L_B = 2\pi/\Omega_L$, then within $L_B/2$, Eq. (3.11) becomes approximately

$$\frac{\Delta\gamma_L}{L_B/2} \approx 2\frac{k_s a_s a_\omega}{\gamma_r} \tag{3.15}$$

or

$$\frac{\Delta\gamma_L}{\gamma} \approx \frac{L_B}{2}\left(\frac{k_s}{\gamma_r^2}\right)(2a_s a_\omega) = 2k_0 a_s a_\omega L_B \tag{3.16}$$

Returning to Eq. (3.8) for $d\psi/dz$, if ψ changes by π, then $\Delta\gamma$ becomes $\Delta\gamma_L$ and $\Delta z = L_B/2$; then

$$\frac{\pi}{L_B/2} \approx 2k_0\left(\frac{\Delta\gamma_L}{\gamma}\right) \tag{3.17}$$

Combining Eqs. (3.16) and (3.17), we get

$$\frac{\Delta\gamma_L}{\gamma} \approx 2\sqrt{a_s a_\omega} \tag{3.18}$$

(Here we have taken the liberty of setting $\sqrt{2\pi} \approx 2$.) Now return to Eq.

(3.14), and set $L = L_B/2$, where L is the undulator length; this assures that the electrons are kicked out of the FEL after the maximum energy extraction:

$$L_B = 2L = \frac{l_0}{2\sqrt{a_s a_\omega}} = \frac{l_0}{\Delta\gamma_L/\gamma} \tag{3.19}$$

where we have used Eq. (3.18); hence $\Delta\gamma_L/\gamma \approx l_0/2L = 1/2N$. The efficiency, η is then

$$\eta = \frac{\Delta W}{W} = \frac{\Delta\gamma_L mc^2}{(\gamma - 1)mc^2} \approx \frac{\Delta\gamma_L}{\gamma} = \frac{1}{2N} \tag{3.20}$$

This expression tells us the maximum energy that can be extracted from an electron as it passes through the undulator. Clearly, since it is required that N be a rather large number on grounds of gain, the FEL efficiency is not high.

The separation of the peaks of positive and negative gain in Fig. 3.4 is $(\gamma_+ - \gamma_-)/\gamma \approx 1/N$ according to Eq. (3.20). Referring to Eq. (2.17), we observe that if the part of the light wave which is moving exactly down the FEL axis finds itself at maximum gain, then a portion of the light which is moving slightly off axis by an amount $\Delta\theta_0 \approx 1/\gamma N^{1/2}$ will interact with the electron beam at the point of absorption. Since $\gamma \sim 100$ and $N^{1/2} \sim 10$, $\Delta\theta_0 \sim 10^{-3}$. To avoid this complicated interaction, one may aperture the FEL or choose a mirror of suitably small radius.

Since γ_{inj} changes by $\Delta\gamma_L$, we expect that the radiation energy emerging from the FEL will not be entirely monochromatic. This can be appreciated by inspecting Fig. 3.4: as the gain drops from the maximum to near zero, the abscissa must also change by $\Delta\omega_L$. From the FEL relationship we know $\Delta\omega_L/\omega_s \approx 2\Delta\gamma_L/\gamma$; hence $\Delta\omega_L/\omega \approx 1/N$. The coherence of the radiation is inversely proportional to the number of undulator periods. Thus one cannot solve the problem of low efficiency by reducing the number of field ripples —usually N is a number in the range 50–100. This is also that one might qualitatively expect from the argument that the electron is asked to scatter N periods of undulator virtual photons. (The spectral width of radiation emitted from an FEL oscillator is narrower than this, because of the high-Q Fabry-Perot resonator.)

A monoenergetic beam of electrons will exit the undulator with an energy spread of order $1/2N$, because these electrons must be injected into the ponderomotive wave with a spread of of initial phases. Each electron loses a somewhat different amount of energy to the wave, and the ensemble exits the undulator with a broadened energy distribution. An important small-signal theorem, derived by Madey et al. [127] and refined by Kroll et al. [108], relates the net energy loss of the electrons, $\langle\gamma_f - \gamma_i\rangle$, averaged over the entry phases into the undulator, to the phase-averaged energy spread $\langle (\gamma_f - \gamma_i)^2 \rangle$:

$$\langle \gamma_f - \gamma_i \rangle = \frac{1}{2} \frac{\partial}{\partial \gamma_i} \langle (\gamma_f - \gamma_i)^2 \rangle \qquad (3.21)$$

The average energy loss is proportional to the gain in optical power, and the spontaneous emission power is proportional to $\langle (\gamma_f - \gamma_i)^2 \rangle$. The theorem connects the mechanism of FEL gain with the broadening of the electron energy distribution. Hence, if an electron beam should be "recycled" through the undulator, it must be allowed to "cool" before reinjection. In certain FELs where the electron pulses are very short, the energy broadening is even larger and more complicated owing to the way the electron pulse travels with respect to the optical pulse.

Arbitrary Amplitude

Now let us return to the point, at the end of the first subsection, where we branched the discussion according to the size of the laser signal as represented by a_s. To begin, suppose a_s is small. In the pendulum model, this amounts to a weak gravitational field, so the pendulum easily swings completely around; most electron orbits are untrapped (Fig. 3.5). Weak a_s means that Ω_L is small, specifically with respect to the parameter $\Delta \omega$:

$$\omega_B \equiv c\Omega_L \ll \Delta \omega \qquad (3.22)$$

where

$$\Delta \omega \equiv \beta_r k_0 c - \omega_s (1 - \beta_r) \qquad (3.23)$$

We can change Eq. (3.13) to the form

$$\frac{d^2 \psi}{dt^2} = -\omega_b^2 \sin \psi(t) \qquad (3.24)$$

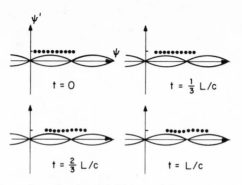

Figure 3.5 Case where the signal field is weak [47]. © 1980 Addison-Wesley.

which has a first integral

$$\left(\frac{d\psi}{dt}\right)^2 = 2\omega_b^2(\cos\psi - \cos\psi_0) + (\Delta\omega)^2 \qquad (3.25)$$

where

$$\psi(t) = \psi_0 + (\Delta\omega)t + k_s[z(t) - z_0]$$

and

$$\psi_0 = k_s z_0 + \phi - \frac{\pi}{2} \qquad (3.26)$$

with ϕ being the phase angle of the optical wave. The quantity $\Delta\omega$ measures how far off resonance [see Eq. (2.25)] the electron is. Then Eq. (3.25) is expanded in the small quantity $(\omega_b/\Delta\omega)^2$ and integrated to get $z(t)$ and $\gamma(t)$ [46]. One considers the expansion initially to terms of order $(\omega_b/\Delta\omega)^2$, and later to order $(\omega_b/\Delta\omega)^4$. The second-order terms show that half the electrons are positioned so that they gain energy and move ahead of the average flow, while the other half slip behind. This causes a spatial bunching, and a symmetrical splitting of the energy distribution into two parts (Fig. 3.6): in one part, work has been done on the electrons by the radiation, while the other group of electrons has done work on the radiation. When the fourth-order terms are included in the expansion, the energy distribution is no longer symmetrical (Fig. 3.6). (An exception occurs if $\Delta\omega = 0$.) The asymmetry corresponds to the entire electron beam having given up net energy to the radiation. It is clear that a substantial reshaping of the electron distribution is involved in the FEL interaction. The skewing of the electron energy distribution has been observed experimentally [60].

To calculate the gain, it is necessary to find the net energy change of all electrons. The result is

$$\frac{\bar{\gamma} - \gamma}{\gamma} = \frac{\gamma_\parallel^2 \omega_b^2}{\omega_s(\Delta\omega)^3}\left[\cos(\Delta\omega t) - 1 + \frac{\Delta\omega t}{2}\sin(\Delta\omega t)\right] \qquad (3.27)$$

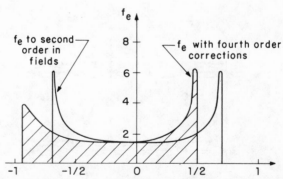

Figure 3.6 Distribution of electron energies in the undulator [46]. © 1977 Addison-Wesley. Abscissa in arbitrary energy units.

It is convenient to measure time as a dimensionless fraction $\hat{t} = t/(L/\beta c)$, which is just the fraction of time an electron spends in the undulator. Fixing L, the parameter to be varied is $L\Delta\omega/\beta c$. As $\hat{t}$ increases, the gain changes, depending on the choice for $\Delta\omega$. For $\Delta\omega = 0$, the gain is zero. The gain is optimized for a certain choice of $\Delta\omega$; as $\Delta\omega$ continues to increase, the gain becomes an oscillatory function along the undulator (Fig. 3.7). We see that the two-wave FEL gain process is an interference effect which is dependent upon the length of the undulator. Even under optimum conditions, note that the end of the undulator is the most important part with respect to the rapid increase of the signal. Our aim is of course to compute the gain G at $\hat{t} = 1$ (the end of the undulator). Recasting Eq. (3.27) in the form of the energy gain, and combining the several terms in the parenthesis, we have

$$ G = G_0 \left[-\frac{1}{2} \frac{d}{dx} \left(\frac{\sin^2 x}{x^2} \right) \right] \Bigg|_{x = [(\Delta\omega) L / 2\beta c]} \tag{3.28} $$

where we have set $\hat{t} = 1$; the gain is found—remarkably enough—to be proportional to the slope of the spontaneous-emission line shape [Eq. (2.38)] and is shown in Fig. 3.8. This was already mentioned in Chapter 2 in connection with the Madey theorem. By examining the coefficient in the brackets, one finds that there is an optimum choice for $\Delta\omega$, or rather a

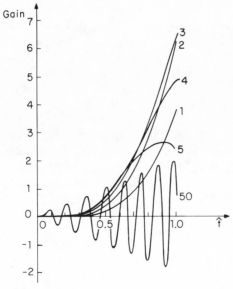

Figure 3.7 Signal gain along the undulator: varying frequency mismatch [46]. The parameter characterizing the curves is $L\Delta\omega/\beta c$. © 1977 Addison-Wesley.

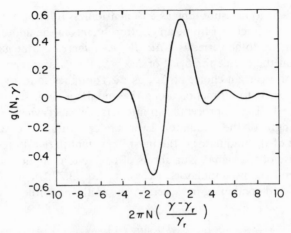

Figure 3.8 FEL gain (quantitative picture).

combination of undulator length and initial energy, given by

$$N = 0.207 \frac{\gamma_r}{\gamma - \gamma_r} \tag{3.29a}$$

or

$$(\Delta\omega)_{opt} = 2.61 \frac{\beta c}{L} \tag{3.29b}$$

for which the gain function $g(x) = -(d/dx)[(\sin^2 x)/x^2]$ is maximized (at $x = 1.303$); at this value, the maximum of g is 0.54, and then G is given by the function

$$G_N = 0.27 \left(ne^4 B_\perp^2 l_0 \right) \left(\frac{Nl_0}{\gamma mc^2} \right)^3 \alpha \lambda_s^{\frac{3}{2}} \tag{3.30}$$

In applications where the electron beam is appreciably smaller than the cross section of the radiation beam with which it interacts, it is important to include a "filling factor" in Eq. (3.30)—this is the ratio of the electron-beam area to that of the radiation. In the Stanford oscillator experiment, the filling factor was 0.017, and the gain is reduced by this factor. Higher maximum gain can be obtained with a longer undulator and larger transverse field, within certain physical constraints that will be discussed later in this book. Note that $(\Delta\omega)_{opt} \propto 1/L$ and therefore the FEL gain vanishes in the limit of infinite undulator length: this shows that there can be no "population inversion" analogy in the two-wave FEL.

The gain in Eq. (3.30) is appropriate to the helical undulator. For a linear undulator, particularly at large a_ω, this formula must be multiplied by $[J_0(\zeta) - J_1(\zeta)]^2$ where $\zeta = (a_\omega^2/4)(1 + a_\omega^2/2)$. As was noted in Eq. (2.21) in Chapter 2, the linearly polarized undulator emits odd-integer harmonic spontaneous radiation along the axis. From the Madey theorem, the gain should be proportional to the derivative of the spontaneous spectrum, and therefore one might expect to find gain at the harmonics as well. [The multiplying coefficient becomes $n\{J_{(n-1)/2}(n\zeta) - J_{(n+1)/2}(n\zeta)\}^2$].This provides an attractive possibility for generating shorter wavelengths using the same electron energy. There are also the penalties one might expect: the gain decreases with increasing harmonic number, and the requirements on electron-beam quality [see Sections 4.4 and 5.5] are more stringent.

The parameter $\omega_b^2 L^2/c^2$, the dimensionless coefficient on the right-hand side of the pendulum equation, governs the type of physics. If it is $< \pi$, the model shown in Fig. 3.5 is appropriate; if it is $> \pi$, consult Fig. 3.3. If $\omega_b^2 L^2/c^2 \gg 1$, the FEL becomes saturated; for $\gamma = 50$, $B_\perp = 2.5$ kG, this occurs at $P \sim 10^8$ W/cm^2. To understand how this happens, remember that the pendulum equation involves the amplitude and phase of the radiation field. As the latter grows, it becomes important to solve the electron dynamics and the radiation self-consistently. The wave equation for the vector potential of the radiation is excited by the transverse current $enca_\omega/\gamma$; it can be solved under the approximation of slow variation of the wave amplitude and phase [see Section 4.2 for a similar calculation]. Two first-order wave equations result for the radiation amplitude and phase [47]:

$$\dot{a}_s = A\left\langle \frac{\sin\psi}{\gamma} \right\rangle$$

$$a_s\dot{\phi} = A\left\langle \frac{\cos\psi}{\gamma} \right\rangle \tag{3.31}$$

where the average $\langle \cdots \rangle$ is made over all the electrons—injected into the undulator with a random distribution of initial phases—and we define $A = a_\omega\omega_{p0}^2 L/2c^2k_s$ and $\dot{a}_s = da_s/d\hat{\imath}$. [For the linear undulator at large a_ω, A should be multiplied by $\{J_0(\zeta) - J_1(\zeta)\}^2$ and the coefficient on the right-hand side of the pendulum equation should be multiplied by $\{J_0(\zeta) - J_1(\zeta)\}$.] Initially, $\langle \cdots \rangle = 0$, but bunching permits these averages to accumulate nonzero values, thereby driving a_s. Colson and Freedman [51] have solved the pendulum equation together with Eq. (3.31) numerically and find that in a strong EM field the gain decreases (Fig. 3.9), due to degradation of the bunching from large changes in the phase function. Note that at high optical power the spectral width of the gain curve also increases. In an oscillator, as the signal builds up in the cavity, it is easy to understand how this will reduce the gain until the system approaches equilibrium—the residual gain just compensating for cavity losses at saturated power.

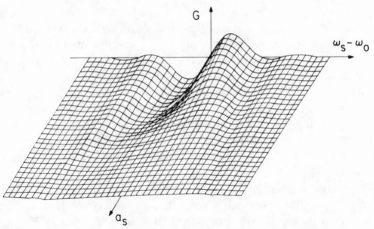

Figure 3.9 Effect of signal amplitude on FEL gain [51]: result for low-electron-density (low-gain) case. © 1983 APS.

Another consequence of strong optical (EM) field amplitude is the possible development of the "sideband" or "synchrotron" instability, in which power appears at frequencies not at the FEL fundamental. This instability is caused by a coupling of the radiation field in the FEL to the synchrotron oscillations of the electrons, but it does not grow until the system is near saturation [70]. These sidebands, separated from the FEL carrier by a fractional frequency shift $\delta\omega/\omega \sim \pm 1/N$, may grow to contain as much as half the total radiated power. This clearly reduces the usefulness of the FEL as a coherent light source, yet it can produce more power as a result of these sidebands. The sideband instability should be controllable by introduction of a dispersive element into the resonator. Under certain conditions, the sideband instability may modulate the short optical pulses emerging from the FEL into a sequence of even shorter pulses.

3.4 Nonuniform-Undulator Effects

Tapered Undulator

The treatment under "Large Amplitude—Approximate Result" in Section 3.3 suggests how one might improve the FEL efficiency. If electrons can be trapped into a ponderomotive well or "bucket" moving at nearly the same speed, under appropriate circumstances it is possible to "lower the center of gravity" of the bucket adiabatically (Fig. 3.10a). Figure 3.10b shows the modified phase diagram; the shaded area within the separatrix contains the trapped electrons. The electrons will "slosh" in the bucket at the synchro-

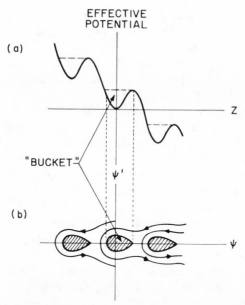

Figure 3.10 Phase diagram for tapered undulator; ponderomotive potential above.

tron frequency, but the energy of the trapped electrons can be reduced on the average. In this manner far more energy may be extracted from the electron stream than the amount predicted by Eq. (3.20). However, the character of the undulator must be varied adiabatically if this is to happen, since the electron energy is no longer constant, even in the average sense. If we return to Eq. (2.17), the FEL equation, and realize that ω_s is a constant during the FEL interaction, we can arrange to change γ_r by means of a corresponding change in k_0 and/or a_w:

$$\gamma_r(z) = \left[\frac{k_s}{2k_0(z)} \left[1 + a_\omega^2(z) \right] \right]^{1/2} \tag{3.32}$$

The prescription is to follow the conversion of electron kinetic energy into radiation, i.e., $\Delta\gamma_r = \gamma_r(0) - \gamma_r(z)$, and "program" the undulator period and/or amplitude so that it matches this change. This can be done by adiabatically changing the undulator period, the amplitude $B_\perp$, or both: this aspect of the problem is at the discretion of the designer.

Before one can start the buckets moving down the "staircase" (Fig. 3.10a), it is necessary to trap the electrons in the bucket. To trap the electrons, however, it is necessary to have a large product for $a_s a_\omega$, as shown by Eq. (3.18); hence, if a_s is not large enough, a_ω must be increased adiabatically while the undulator period is changed so as to keep the FEL frequency and resonance energy fixed, according to Eq. (3.32).

This rather sophisticated problem can be solved using the method outlined in Section 3.3 under "Arbitrary Amplitude," but there are a few modifications that must be made, including a familiar and valuable approximation. First, the "constant torque" term on the right side of the pendulum equation (3.13) must be retained. For example one can show that the size of this term depends on the undulator period nonuniformity, or "taper," $\Delta l_0 / l_0$; this is also what causes the stair-step modification in the ponderomotive potential (Fig. 3.10a) as well as the distortion of the separatrix surfaces (Fig. 3.10b) which enclose the electrons trapped in the "bucket." Next, one should note that the k_0 and a_ω terms in Eq. (2.31) now vary slowly and regularly with z. The approximation referred to involves a well-known technique used in accelerator physics, wherein we replace the dynamics of the electrons in the bucket by a representative electron which moves in resonance with the ponderomotive well. If a partly filled bucket of electrons is sent down a properly designed undulator, it will respond the same way as the resonant electron does.

The problem is formulated in the following way. First, a slow (adiabatic) variation of the undulator period and/or amplitude is chosen. The equations (3.31) are used, but the averages now will pertain to the resonant electron: e.g., $de_s/dz \propto \sin \psi_r$. Equation (3.9) is used to find the change of energy of the resonant electron,

$$\frac{d\gamma_r}{dz} = - \frac{k_s a_s a_\omega}{\gamma_r} \sin \psi_r \qquad (3.33)$$

We observe that $\psi_r > 0$ corresponds to electron deceleration and increase in e_s, whereas $\psi_r < 0$ describes acceleration and decrease in e_s. From this is obtained the energy inequality

$$\gamma_r^2(0) - \gamma_r^2(L) < 2k_s a_s a_\omega L \qquad (3.34)$$

The resonant electron can be moved along the undulator by requiring that it remain in phase with the ponderomotive well, i.e., $d\psi_r/dz = 0$. For this, we use Eq. (2.31), but take into account the additional effect of the amplitude of the scattered wave [156], which can become appreciable:

$$0 = k_0 - \frac{k_s}{2\gamma_r^2} \left[1 + a_\omega^2 - \frac{2e_s b_\omega}{k_s k_0} \cos \psi_r + \left(\frac{e_s}{k_s} \right)^2 \right] + \frac{d\phi}{dz} \qquad (3.35)$$

Although computational studies are generally made to find the amplification, Slater [176] has estimated the extraction efficiency for small decrease in electron energy:

$$\eta \approx \left(\frac{n_t}{n} \right) \frac{a_\omega \sin \psi_r}{\gamma_r^2} \int_0^L e_s(z)\, dz \qquad (3.36)$$

which shows that an optimization of the fraction of electrons trapped in the bucket (n_t/n) and the resonant phase ψ_r is necessary. The linewidth of an FEL with a linear tapered undulator is $\sim (l_0/L)^{1/2}(\Delta\gamma_r/\gamma_r)^{1/2}$, where $\Delta\gamma_r$ is the resonant-energy change along the undulator [41]; $\Delta\gamma_r$ and the taper are related by eq. 3.32.

As the ponderomotive buckets fall in energy, some of the trapped electrons will spill out: the faster the fall, the fewer electrons retained (a more quantitative statement is that n_t/n decreases with increasing ψ_r [108]). An optimization for ψ_r and n_t [40] shows that the actual situation is very similar to that encountered in particle accelerators. About 40% of the electrons can be trapped, and the energy of this group can be decreased about 15%, giving a net extraction efficiency of roughly 6%. This is a very attractive enhancement of efficiency—an order-of-magnitude improvement over the uniform undulator. Because of the dependence of diffraction on wavelength, the result is independent of laser wavelength.

The variation in undulator period or amplitude must be adiabatic; otherwise electrons will spill out of the buckets. Ideally, this would require a very long undulator. This is not only expensive, but rather more importantly the lengthening of the undulator raises the question of diffraction. Practical electron beams have finite radial extension r_b, and an optical beam can be focused to this dimension only for a limited axial distance of order of the Rayleigh range, $L_R = \pi r_b^2/\lambda_s$. Since we should take $L_R < L$, the adiabatic tapering of the undulator period can occur only in some practical limit. Furthermore, the shape of the optical pulse that is amplified is usually described with a Gaussian radial profile, and this raises the question of two-dimensional effects, such as focusing, or the effect of radial variation of a_s or a_ω upon trapping. These are important complications which have been treated theoretically and computationally [108, 191].

The penalty for efficiency enhancement is that the gain of the FEL must be low. The efficiency-enhanced FEL requires a high-power signal wave (which can combine with the undulator to yield a large $a_s a_\omega$) to trap the electrons. The small-signal gain is low because the electrons remain near resonance for only a short distance near the undulator entrance. On the other hand, the requirements on the electron-beam energy spread are liberal. Another consequence of high power—which follows from the efficiency enhancement technique—is the possible development of additional spectral components due to the "sideband instability" [6]. There may be difficulty in maintaining a high degree of trapping as the power becomes very large, due to the presence of sideband instabilites as well as the finite radial structure of the beam.

A very similar technique for efficiency enhancement involves the use of an empty bucket. The empty bucket can be accelerated through a broad electron momentum distribution in phase space using an undulator with a "reverse" tapered period. As the bucket rises through the distribution, electrons passing around the bucket give up energy to the optical field. In this method of "phase-space displacement" [108], none of the electrons are trapped.

We shall return to the important subject of the variable-parameter undulator again theoretically in Section 4.5, where the one-dimensional formulation is presented in detail, including space-charge effects on the beam.

Optical Klystron

Another modification of the FEL which has rather the opposite purpose to a variable-parameter undulator is the *optical klystron*, which allows for high gain in weak optical fields but low gain in strong optical fields [50]. First described by Vinokurov and Skrinsky at Novosibirsk, the optical-klystron modification is potentially useful in FELs at very short (optical) wavelength such as might be associated with a high-energy storage ring, where the gain is so low that the laser might not oscillate were the gain not enhanced. In the optical klystron the undulator is separated into two parts by a "dispersive" region. The purpose of this separation is to permit the electrons to fall behind the optical wave in an energy-dependent way. This can be done either with empty space, or—to use the total space more efficiently—by introducing a bend in the electron trajectory which temporarily deflects the electrons away from the magnetic axis. In either case, a small spread in electron energy established in the first undulator section can cause electron phase shifts to bunch electrons in the second undulator section. The first section pre-bunches the beam, permitting us to retain in the second section the desirable portion of the system where most of the gain occurs (see Fig. 3.7).

The gain can be derived from the spontaneous spectrum, using Madey's theorem. This spectrum [3] involves an integration over the electron trajectory which is broken into three parts—the zones of the two undulators (N periods each) separated by the short dispersive section. The latter introduces a phase factor $2\pi N_d$ into the result, where N_d is the number of optical wavelengths passing over the electron in the dispersive section [68]. If $N_d \gg N$, it is found that the optical-klystron gain is about $1 + N_d/N$ times as great as the gain of a uniform undulator having $2N$ periods. The penalty for obtaining higher gain using the optical klystron modification to the FEL is that one requires a higher-quality (lower-emittance) electron beam. The optimal N_d that maximizes the optical klystron gain is of order $1/(4\pi\,\delta\gamma/\gamma)_{\parallel}$, where $(\delta\gamma/\gamma)_{\parallel}$ is the normalized beam parallel-momentum spread. The optical klystron can also be studied using a numerical approach to the pendulum equation, by interrupting the interaction with the undulator for a fixed interval at a chosen location (position of the dispersive section).

The first electron storage-ring FEL [29] to oscillate as an FEL required the use of an optical-klystron undulator configuration to increase the gain beyond oscillation threshold.

Transverse-Gradient Undulator

A monoenergetic electron beam is ideal for good FEL operation. On the other hand, many accelerators have unacceptably large energy spread ($> 1/N$), and in the electron storage ring the distribution of electron energies increases after each passage of the electron pulse through the undulator. The transverse-gradient undulator was proposed [180] to accommodate the undesirable beam energy spread by a modification of the undulator.

Suppose the beam electrons can be dispersed spatially, in the coordinate transverse to the beam axis, according to energy. This could be done by a bending magnet. Looking again at the FEL equation (2.17), and noting that the frequency of the entire system should be constant, we observe that if γ varies in the transverse direction [say, as $\gamma(x) = \gamma_0(1 + ax)$], then one can compensate for this variation by an appropriate programming of the undulator amplitude in the transverse direction: $B_\perp(x)$. Electrons with slightly higher energy should encounter the stronger undulator field, and vice versa. The gradient $dB_\perp/dx$ is chosen to match the gradient $d\gamma/dx$ so that λ_s will be the same for all electrons: $(1/\gamma)\,d\gamma/dx = [a_\omega^2/(1 + a_\omega^2)](1/B_\perp)\,dB_\perp/dx$.

When the electron beam has an energy spread $\sim 1/N$, the performance of the transverse gradient undulator will be superior to that of an equivalent uniform undulator, as all electrons will enter with the same resonance frequency. For very similar reasons, the transverse-gradient undulator gain will decrease more slowly with increasing optical (EM) field amplitude, and the saturation of the FEL will be extended to higher power. The transverse-gradient undulator "expands" the gain of the FEL.

At one time, the insensitivity of the transverse-gradient undulator to energy spread was believed to be an attractive solution to the storage-ring FEL problem (Chapter 8). However, Kroll [108] has shown that the transverse-gradient undulator should excite transverse betatron oscillations of the electron beam; these oscillations place a constraint on the operation of the system that amounts to an upper bound on the laser power which may be generated. An important consequence has been that the Madey gain-spread theorem has been generalized for such two-dimensional cases [110]. It has been established that one cannot design an undulator which produces FEL gain and yet causes vanishing energy spread or transverse excitation of the electron beam.

An excellent comparison of the properties of the different nonuniform undulators has been made by Colson and Freedman [51].

This is surely not the end to the list of possibly useful nonuniform undulators. Numerical simulation of the dynamics can handle important combinations of the above, or new varieties adapted to special applications. An example of this sort of design is to be found in the work of the TRW group [174], where a multicomponent undulator has been developed which combines features of the tapered undulator and the optical klystron. Both large- and small-signal gain can be enhanced, which is an attractive possibility for the FEL oscillator, where the power changes very rapidly as the system begins to oscillate. Use of such an undulator was responsible for the successful oscillation of an FEL on the third harmonic, at $\approx 0.5 \ \mu$m, reported by Edighoffer et al. [63].

3.5   Transient Effects: Lethargy and Startup

Thus far we have discussed the FEL under the assumption that the electron beam current is uninterrupted. This is not the case in certain important classes of rf accelerators: the rf linac, the storage ring, and the microtron, where the electrons are accelerated by an external rf field. The current from such an accelerator arrives as a series of short pulses (Fig. 1.6), perhaps as short as 3 psec in duration, or about 1 mm long. Obviously one needs to synchronize the electron pulse with the pulses of amplified light. In FEL oscillators, this is done by choosing the Fabry-Perot cavity length L_c appropriately. The bounce time of radiation in the optical cavity is $2L_c/c$, and this must be nearly equal to the temporal spacing L_p/v of the current pulses from the accelerator (Fig. 3.11). Most of the cavity—except for the volume centered on the electron pulse—contains no radiation.

At the end of N periods of the undulator, the resonance relation shows that the light has slipped ahead of the electrons by $N\lambda_s$. When this is comparable to the length of the current pulse, l_p, short-pulse effects will be

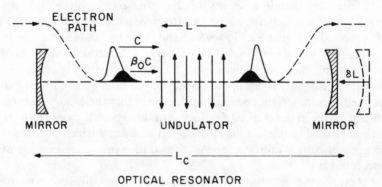

Figure 3.11   Electron and optical pulses in an FEL.

observed. The ratio $s = N\lambda_s/l_p$ is a number of order unity for some FEL experiments which use rf accelerators. The effect of the short pulse length is that the FEL output power becomes extremely sensitive to the resonator length. The time scale of the slippage, characterized by s, is of the same order as the response of the FEL gain medium (where the fractional bandwidth is $\sim 1/N$), and this causes gain to occur on the trailing edge of the light pulse. This effect is known as *laser lethargy*. The concept of the lethargy was developed in connection with X-ray laser amplifiers [98], but the first application was to the FEL [16]. Several groups have refined the analysis (e.g., [47, 186]) and have removed limitations present in the original model due to its connection with atomic-laser theory.

The bunch length of electrons emerging from rf-type accelerators also has an effect on the coherence. The length of the radiation pulse is constrained by the length of the current pulse, l_p. The radiation line width, to the extent that it is created by this effect, is given by

$$\delta\omega \sim \frac{\pi c}{l_p}, \qquad \text{or} \qquad \frac{\delta\omega}{\omega} \sim \frac{\lambda_s}{l_p} \qquad (3.37)$$

which can be different from the estimate (for the DC electron beam) given by Eq. (3.21). For $l_p = 1$ mm and $\lambda_s = 1$–10 μm, then $\delta\omega/\omega$ is in the range 0.1–1.0% for a typical FEL operating in the IR.

We can understand the lethargy effect by considering the following argument. Suppose we have adjusted the resonator of the FEL so that $L_p/v = 2L_c/c$, so we expect the bounce time of the photons in the resonator will just synchronize with the electron pulses. However, there is more gain at the end of the undulator than at the beginning (Fig. 3.7) so the trailing edge of the optical pulse is amplified more than the leading edge. After several reflections from the mirrors, the centroid of the optical pulse lags behind the electrons, and it will intercept the next electron pulse later than $2L_c/c$. If nothing is done, eventually the centroid of the light pulse walks off the gain medium, and the system may not generate enough gain to overcome the resonator losses. The "cure" is to reduce L_c by a small amount δL to compensate for the slippage.

It has been observed [60] that the optimization of the short-pulse FEL output is extremely sensitive to the quantity

$$\delta L \equiv L_c - \frac{L_p}{2\beta} \qquad (3.38)$$

and that in practice $\delta L \sim 10$ μm was required. Observations also show that both the FEL linewidth and shape, as well as the electron energy spectrum, are sensitive to the parameter δL.

A comprehensive theory of the startup of a pulsed-beam FEL has been proposed by Sprangle, Tang, and Bernstein [186]. The theory takes into

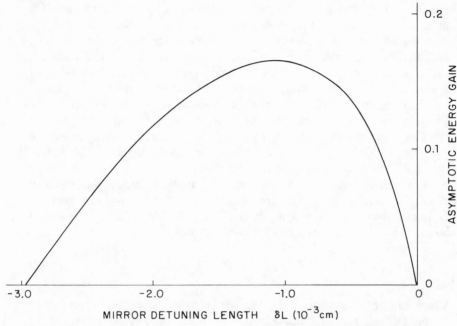

Figure 3.12 Dependence of gain upon resonator detuning [186]. © 1983 APS.

account the fact that the electrons are discrete and uncorrelated. The theoretical prediction (Fig. 3.12) shows how the gain depends on δL. (When comparing with experiment, note that optimum gain does not correspond to optimum power in the oscillator.) Effects of finite beam-pulse length are found to account for a reduction in FEL gain, in addition to that due to the filling factor: in the example of the Stanford FEL, the reduction was $\approx 60\%$.

When mirrors are added to the configuration, the signal regenerates. Suppose the average net gain per bounce of radiation in the optical resonator (including the FEL gain with the cavity losses) is $\overline{G}$. Assuming $\overline{G}$ does not change as the signal grows—which is only true for small signals—after M bounces of the radiation between the mirrors we have

$$\frac{P(M)}{P_0} = M - 1 + (1 + \overline{G})^M \approx M + e^{\overline{G}M} \qquad (3.39)$$

provided $\overline{G} \ll 1$ and $M \gg 1$. If $M = 500$, $\overline{G} = 0.06$, and $P = 5 \times 10^{-2}$ W (spontaneous power level), then $P(M) = 30$ MW inside the cavity.

It was observed that the two-wave FEL startup was very long—several hundred bounces—and that this was also sensitive to δL, as was the short-pulse or lethargy effect [60]. The theory by Sprangle et al. [186] can account for these complications, with a simple limit in the long-beam-pulse case. The formulation is one-dimensional, using randomly located charge sheets to model the uncorrelated position of the electrons in the electron beam. The time-dependent current is separated into a coherent $[J_{coh}(z,t)]$ and incoherent $[J_{incoh}(z,t)]$ term, where

$$J_{coh} = -|e|v_\perp f_c \langle n(z,t) \rangle$$
$$J_{incoh} = -|e|v_\perp f_i \big[n(z,t) - \langle n(z,t) \rangle \big] \qquad (3.40)$$

Here f_c and f_i are the coherent-wave and incoherent-wave filling factors (not necessarily the same, because of the resonator geometry). The spatial average over the initial density distribution is represented by $\langle n \rangle$. The EM wave equation is driven by these source terms:

$$\left(\frac{\partial^2}{\partial z^2} - \frac{1}{c^2} \frac{\partial^2}{\partial t^2} - \frac{v_L}{c^2} \frac{\partial}{\partial t} \right) A_s(z,t) = -\frac{4\pi}{c} (J_{coh} + J_{incoh}) \quad (3.41)$$

The losses in the resonator are modeled as a loss rate $v_L = \omega_s/Q$, where Q refers to the resonator Q. The radiation field is spatially Fourier-analyzed into modes $k_q = q\pi/L_c$ which oscillate as $\exp(i\omega_q t)$ with amplitude $a_q(t)$. The calculation is restricted to the small-signal regime, with low gain per bounce. The longitudinal particle dynamics are governed principally by the ponderomotive force.

A lengthy calculation yields the spontaneous-radiation source term $S_{qp}(t)$ and the gain matrix $G_{qp}(t)$. These drive the energy-density matrix $W_{qp}(t) = \langle b_q(t)b_p^*(t) \rangle$ in a radiation-rate equation. In the limit of a long electron pulse (i.e., $l_p \lesssim L_c$), the gain and source matrices are diagonal, and single-mode analysis suffices—laser lethargy is then unimportant. The reduced energy-rate equation becomes

$$\frac{\partial}{\partial t} W_{qq} = \left(\frac{g_{qq}}{\Delta t} - v_L \right) W_{qq}(t) + \frac{S_{qq}}{\Delta t} \qquad (3.42)$$

where $\Delta t = 2L_p/v$, and where

$$S_{qq} \propto \left(\frac{\sin x}{x} \right)^2 \quad \text{and} \quad g_{qq} \propto \frac{d}{dx} \left(\frac{\sin x}{x} \right)^2$$

(x is proportional to the amount of time the electron spends in the undulator). The FEL oscillator power in the resonator evolves as

$$W_{qq}(t) = \frac{S_{qq}}{g_{qq} - v_L \Delta t} \left\{ e^{(g_{qq} - v_L \Delta t)t/\Delta t} - 1 \right\} \qquad (3.43)$$

For parameters appropriate to the Stanford FEL [Table 8.2, Chapter 8], the dependence of the laser power upon the number of light-beam passes

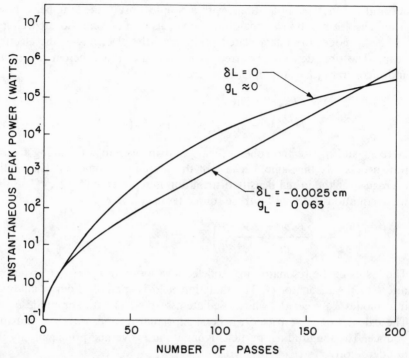

Figure 3.13 Startup of the Stanford FEL oscillator, calculation by Sprangle, Tang, and Bernstein [186]; g_L = small-signal gain. © 1983 APS.

has been calculated, with δL as a parameter (Fig. 3.13); note that this case does not correspond to Eq. (3.43), and the off-diagonal terms in the energy rate equation are important. It is apparant that not only does this FEL require ~ 100–200 passes to approach steady state, but also the resonator detuning δL is very important as well. The actual FEL signal is shown in Fig. 8.5, and agrees well with the calculation outlined above. The startup of an FEL oscillator is similar to that of a classical oscillator, or for that matter, its cousin the atomic laser [115].

4

FEL Theory, Part II:
Dense-Electron-Beam
Effects

4.1 *Introduction*

The treatment of Chapter 3 omits potentially important effects which can arise from consideration of space charge (Section 2.6). All interactions in Chapter 3 occurred as the result of individual electrons interacting with the undulator and the laser field. This interaction scaled as n, or ω_p^2, and it might appear that increasing the beam density would provide the benefit of higher gain and higher power. Indeed this does happen in a qualitative way, but one must reformulate the theory to properly understand what is involved. When this is done one finds that sometimes gain is increased, sometimes reduced, by raising n. A new variable is added to the system, that is, there is another repository for energy: in the electrostatic field of the charge bunches.

We begin this chapter with an elementary treatment of the FEL in the limiting case of high charge density, where the electrostatic energy resides in a plasma or space-charge wave. The model we use is the one of three-wave parametric pumping, familiar to readers versed in microwave devices or parametric lasers; it was first applied to the FEL by McDermott [204]. In this approach, we obtain quantitative information about the operation of the FEL in the convective-instability "collective" mode, where the scattered wave will grow exponentially along the undulator. This result is so different from what we have understood from Chapter 3 that one is naturally inclined to seek a more comprehensive model (Section 4.3) that will encompass both two- and three-wave FELs. We conclude the chapter by returning to a generalization of the pendulum equation suitable for systems of arbitrary charge density.

In this chapter we retain our very simplified model of electron motion in the undulator. However, it is fairly obvious that detailed understanding of the FEL is only as sound as is our knowledge of the individual electron orbits. One place where this matters is when the electrons in the beam have a distribution of electron velocities. We shall touch on this problem briefly in this chapter and return for a more detailed treatment in Chapter 5.

4.2   *Parametric-Amplifier Model*

In this section we shall derive the gain for an FEL under the condition known as *stimulated Raman backscattering*, where the beam is cold and dense. As outlined in Section 2.6, the three waves involved are the undulator (in the rest frame), the scattered wave, and the plasma wave. The first two are EM waves, while the third is ES—driven by the ponderomotive force at a wavelength which is approximately $l_0/2\gamma$ in the rest frame. The relationships between frequency and wavenumber have been derived in Chapter 2 and we accept that development as a basis for what follows. In this section, as well as the following, all calculations are done in the rest frame of the electrons, and for simplicity we shall drop the prime in the notation for the calculations of these two sections only. Where we calculate the gain, a scalar, the result will be correct for the laboratory frame, provided the variables are calculated in the same frame. If the growth rate q or distance is calculated, then this result must be Lorentz-transformed to the laboratory frame: Hasegawa [96] has shown that the relation between the two frames is

$$g_{(\text{lab})} \approx \frac{g'_{(\text{rest})}}{2\gamma} \tag{4.1}$$

We begin with Maxwell's wave equation for the field amplitude:

$$c^2 \nabla \times (\nabla \times E) + \frac{\partial^2 E}{\partial t^2} = -4\pi \frac{\partial J}{\partial t} \tag{4.2}$$

The first term on the left-hand side is also represented as $\nabla(\nabla \cdot E) - \nabla^2 E$. Taking the orientation where the scattered wave moves to the right while the undulator field moves to the left, we represent the three waves by

$$\hat{z}\left[E(\omega_i, t)e^{i(\omega_i t + k_i z)} + \text{c.c.}\right], \qquad \text{idler (plasma)} \qquad (4.3)$$

$$\hat{x}\left[E(\omega_s, t)e^{i(\omega_s t - k_s z)} + \text{c.c.}\right], \qquad \text{signal (scattered)} \qquad (4.4)$$

$$\hat{x}\left[E(\omega_0, t)e^{i(\omega_0 t + k_0 z)} + \text{c.c.}\right], \qquad \text{pump (undulator)} \qquad (4.5)$$

where $\omega_0 = 2\pi\gamma v/l_0$, $\omega_s = 2\pi\gamma v/l_0 - \omega_p$, $k_0 = 2\pi\gamma/l_0$, $\omega_s = k_s c$, and $\omega_i \approx \omega_p$. Differentiation of Eq. (4.3) gives

$$\frac{\partial^2}{\partial t^2}\left[E(\omega_i, t)e^{i(\omega_i t + k_i z)}\right] = \left[-\omega_i^2 + 2i\omega_i\frac{d}{dt} + \frac{d^2}{dt^2}\right]E(\omega_i, t) \quad (4.6)$$

Substituting the waves into the left side of (4.2) gives

$$\left(-\omega_i^2 + 2i\omega_i\frac{d}{dt} + \frac{d^2}{dt^2}\right)E(\omega_i, t) = -4\pi i\omega_i J_z(\omega_i, t) \quad (4.7)$$

$$\left(c^2 k_s^2 - \omega_s^2 + 2i\omega_s\frac{d}{dt} + \frac{d^2}{dt^2}\right)E(\omega_s, t) = -4\pi i\omega_s J_x(\omega_s, t) \quad (4.8)$$

$$\left(c^2 k_0^2 - \omega_0^2 + 2i\omega_0\frac{d}{dt} + \frac{d^2}{dt^2}\right)E(\omega_0, t) = -4\pi i\omega_0 J_x(\omega_0, t) \quad (4.9)$$

where the $c^2 k^2$ terms in Eqs. (4.8) and (4.9) come from the $\nabla^2 E$ operation [they do not appear in (4.7) because $\nabla \times E(\omega_i) = 0$]. The current contribution comes from two categories: a linear current J_L and a nonlinear current J_{NL}. The linear current appropriate for Eq. (4.7) is

$$J_L = nev(\omega_i) \qquad (4.10)$$

but $\dot{v} = eE/m$ and $v(\omega_i) = eE/im\omega_i$, so

$$4\pi i\omega_i J_L(\omega_i) = \frac{4\pi ne^2}{m}E(\omega_i, t) \equiv \omega_p^2 E(\omega_i, t) \qquad (4.11)$$

and similarly for Eqs. (4.8) and (4.9). Next, we drop the d^2/dt^2 operation in comparison with $\omega\, d/dt$. This assumes the amplitude grows slowly compared with the oscillation frequency of the wave. Then Eqs. (4.7)–(4.9) become

$$\left(\omega_i^2 - \omega_p^2\right)E_i(\omega_i, t) = 2i\omega_i\left[\frac{dE_i}{dt} + 2\pi J_{z\,\text{NL}}(\omega_i, t)\right] \quad (4.12)$$

$$\left(\omega_s^2 - \omega_p^2 - c^2 k_s^2\right)E_s(\omega_s, t) = 2i\omega_s\left[\frac{dE_s}{dt} + 2\pi J_{x\,\text{NL}}(\omega_s, t)\right] \quad (4.13)$$

$$\left(\omega_0^2 - \omega_p^2 - c^2 k_0^2\right)E_0(\omega_0, t) = 2i\omega_0\left[\frac{dE_0}{dt} + 2\pi J_{x\,\text{NL}}(\omega_0, t)\right] \quad (4.14)$$

where we have used $J = J_L + J_{NL}$. The left-hand side of each of these equations is very nearly zero, because in the absence of growing, coupled linearized waves, $dE/dt = 0 = J_{NL}$ and we are left with the linearized dispersion relation for each mode:

$$D(\omega, k)E = 0 \qquad (4.15)$$

Each wave very nearly satisfies Eq. (4.15); hence the slow growth—if any—proceeds according to

$$\frac{dE}{dt} = -2\pi J_{NL} \qquad (4.16)$$

To keep the analysis simple, we proceed by letting the pump wave have constant amplitude (in reality the undulator pump will set up perturbations on the beam which will deplete the pump wave in the electron frame, despite the fact that the undulator remains unchanged).

The next task is to search for the nonlinear currents. Here, "nonlinearity" means that there are frequencies present in the output which were not present in the input (nonlinear and small-signal are nevertheless compatible terms here). The dominant nonlinear current contribution to $J(\omega_i, t)$ comes from the ponderomotive bunching effect:

$$J_{z\,NL}(\omega_i) = -|e|nv_z^{(2)}(\omega_i) \qquad (4.17)$$

where $v_z^{(2)}(\omega_i)$ is derived from the ponderomotive interaction

$$v_z^{(2)}(\omega_i) = \frac{ie}{m\omega_i c}\left[v_x^{(1)}(\omega_0)B_y^*(\omega_s) + v_x^{(1)*}(\omega_s)B_y(\omega_0)\right] \qquad (4.18)$$

where

$$v_x^{(1)}(\omega_0) = \frac{ie}{m\omega_0}E_x(\omega_0) \qquad (4.19)$$

$$v_x^{(1)}(\omega_s) = \frac{ie}{m\omega_s}E_x(\omega_s) \qquad (4.20)$$

$$v_z^{(1)}(\omega_i) = \frac{ie}{m\omega_i}E_z(\omega_i) \qquad (4.21)$$

We shall associate $B_y(\omega_s)$ with $E_x(\omega_s)$ and $B_y(\omega_0)$ with $E_x(\omega_0)$. Combining the terms, (4.18) becomes

$$v_z^{(2)}(\omega_i) = -\frac{4}{\omega_i k_i}\left(\frac{e^2}{m^2 c^2}\right)E_x(\omega_0)E_x^*(\omega_s) \qquad (4.22)$$

and so Eq. (4.12) becomes

$$\frac{d}{dt}E_z(\omega_i, t) = -\frac{2\omega_p^2 e}{mc^2}\left(\frac{1}{\omega_i k_i}\right)E_x(\omega_0)E_x^*(\omega_s) \qquad (4.23)$$

Nonlinear currents such as $n^{(1)}(\omega_s)v_z^{(1)}(\omega_0)$ vanish because EM modes do not drive a density fluctuation on the beam.

The nonlinear current which drives the signal wave is

$$J_{x\,NL}(\omega_s) = -|e|n^{(1)*}(\omega_i)v_x^{(1)}(\omega_0) \qquad (4.24)$$

where the quantity $v_x^{(1)}(\omega_0)$ is given by (4.19). To obtain $n^{(1)}$, the bunching of beam space charge by the ponderomotive force, we use the equation of continuity:

$$\frac{\partial n}{\partial t} + \nabla \cdot (nv) = 0 \qquad (4.25)$$

or

$$i\omega_i n^{(1)}(\omega_i) = -ink_i v_z^{(1)}(\omega_i)$$
$$= -ink_i\left(\frac{ie}{m\omega_i}\right)E_z(\omega_i) \qquad (4.26)$$

We solve for $n^{(1)}$ and substitute it in Eq. (4.24), so that Eq. (4.13) takes the form of Eq. (4.16):

$$\frac{d}{dt}E_x(\omega_s, t) = -\frac{\omega_p^2 e}{2m}\left(\frac{k_i}{\omega_i^2\omega_0}\right)E_z^*(\omega_i)E_x(\omega_0) \qquad (4.27)$$

If we define

$$g^2 \equiv \frac{\omega_p^4\omega_0}{\omega_i^3}\left(\frac{e}{mc\omega_0}\right)^2 E_x^*(\omega_0)E_x(\omega_0) \qquad (4.28)$$

then, differentiating Eq. (4.27) and combining with (4.23) and (4.28), we obtain

$$\ddot{E}(\omega_s, t) = g^2 E(\omega_s, t)$$
$$\ddot{E}(\omega_i, t) = g^2 E(\omega_i, t) \qquad (4.29)$$

Growth (or damping) occurs if the wave phases satisfy $\phi_0 - \phi_s - \phi_i = \pi$ (or 0) respectively. Now take the initial conditions

$$E(\omega_s, 0) = E_0$$
$$E(\omega_i, 0) = 0 \qquad (4.30)$$

and the solution is

$$E(\omega_s, t) = E_0\cosh gt$$
$$E(\omega_i, t) = E_0\sinh gt \qquad (4.31)$$

i.e., a wave will grow as the electron moves along the undulator, so that, at the end of the undulator, the amplitude is $E_0 \cosh gT$, where

$$T = L/\gamma c \beta \qquad (4.32)$$

Equation (4.28), after setting $\omega_i \approx \omega_p$, gives the rest-frame growth rate g'; it is linear in the quiver velocity part:

$$\frac{eE'_\perp(\omega'_0)}{mc\omega'_0} \approx \frac{eB'_\perp(\omega'_0)}{mc\omega'_0} = \frac{eB_\perp}{mc^2 k_0} \qquad (4.33)$$

Then, transforming g' to the laboratory frame by dividing by 2γ,

$$g = \left[\frac{\omega_p}{4k_0 c\gamma}\right]^{1/2} \frac{eB_\perp}{mc} \qquad (4.34)$$

Note that g is linear in the pump amplitude and varies as $n^{1/4}$. This is the case of stimulated Raman backscattering; observe that the dependence on the beam density is very different from what we have found for the two-wave FEL.

The result just obtained applies only if the pump wave is sufficiently weak so that the growth rate $< \omega_p$. Increasing the pump amplitude beyond this point tends to destroy the identity of the plasma wave. This is no disaster—in fact the gain continues to increase—but the new beam equilibrium includes the strong oscillation driven by the pump as the more important feature. Raman FELs have high gain ($gL/c \gtrsim 1$), but they are also long-wavelength devices which experience large diffraction. As the Rayleigh range is short, the undulator must be short, and so operation of the Raman FEL in the high-gain mode is crucial.

Now we have found that not all the frequencies are real quantities, that is, $\omega_0 - \omega_s - \omega_i \neq 0$. The appearance of the imaginary part $\text{Im}\,\omega \approx g$ implies the existence of a comparable real part that "detunes" the system from the previous frequency and wavenumber selection rules. An approximate relation is

$$\text{Re}[\omega_0 - \omega_i - \omega_s] = \Delta\omega_s \approx g \qquad (4.35)$$

which shows that the fractional linewidth will be $\Delta\omega_s/\omega_s \approx g/\omega_s$ (also the fractional linewidth in the laboratory frame), as indicated in Fig. 4.1. We rewrite Eq. (4.35) in a form which recovers $\omega = \omega_s$ for $g = 0$ in the rest frame:

$$\omega - \gamma k_0 \beta c + \omega_p \approx -g \qquad (4.36)$$

Consider the possibility that the scattered and space-charge waves grow until the system saturates, viz., $g = 0$; this results in practice from the

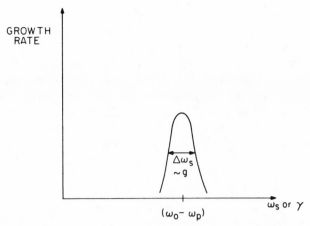

Figure 4.1 Dispersive growth rate of a Raman FEL.

decrease in β (to a value β_{sat}) during signal growth. This decrease is large enough to shift the pump frequency outside the spectrum of unstable growth (Fig. 4.1). Let us suppose the pump remains undepleted while this takes place. When the system is saturated, Eq. (4.36) becomes

$$\omega - \gamma_s k_0 \beta_{\text{sat}} c + \omega_p \approx 0 \tag{4.37}$$

Subtracting (4.37) from (4.36),

$$\gamma k_0 c (\beta - \beta_{\text{sat}}) \approx g$$

or

$$|\Delta\beta| \approx \frac{g}{\gamma k_0 c} = \frac{g}{\omega_s} \tag{4.38}$$

As will be shown in Section 4.4 [Eq. (4.66)], $\Delta\beta$ (in the rest frame) is just $\Delta\gamma/\gamma$, which is related to the efficiency

$$\eta \equiv \frac{\text{change in electron kinetic energy}}{\text{initial kinetic energy}} = \frac{\Delta\gamma(mc^2)}{(\gamma - 1)mc^2}$$

Hence

$$\eta \approx \frac{g}{\omega_s} = \frac{g}{\gamma k_0 c} \tag{4.39}$$

The scaling of η is clearly unfavorable with respect to increasing frequency or γ. Weak-pump Raman theory is valid for $g < \omega_p$, so the maximum Raman efficiency is $\lesssim \omega_p/\gamma k_0 c$.

A mechanism for limiting the output of the Raman FEL is that the slow space-charge wave idler eventually grows large enough to trap electrons [121]. This modifies the distribution of electron energies, which ideally should be monoenergetic. The saturation process is sensitive to the assumption of the initial width of the electron energy distribution, but a qualitative conclusion is that the FEL output will be improved if the beam is originally cold. Saturation is accompanied by many sidebands [122]. The efficiency of the Raman FEL at the point where trapping occurs is ~ $\omega_p/\gamma k_0 c$. A numerical simulation of FEL wave growth and saturation (Fig. 4.2) has been made by Kwan [113]. By following the changes in the electron distribution function, he found that a significant fraction of the electrons were trapped by the electrostatic potential of the idler.

The flow of energy in this "parametric pumping" process can be understood by consulting Fig. 4.3, a Manley-Rowe or energy-level diagram. In Fig. 4.3a, the Stokes process, a pump photon decays into a plasmon and a scattered photon; this is permissible energetically. However, in Fig. 4.3b the anti-Stokes process is not energetically possible unless the plasmon is provided externally (this is usually not possible). Nevertheless, if photons at ω_s are present, we could decrease ω_0 by reducing β, in which case the scattered photon would decay into a plasmon and a pump photon. This is

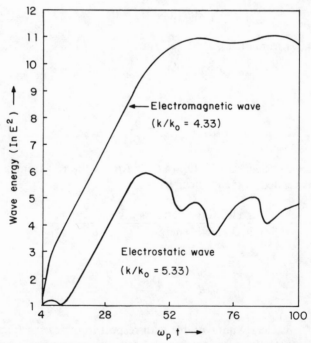

Figure 4.2 Growth of electromagnetic and electrostatic waves, showing saturation, for $\gamma = 2.0$ and $\Omega_\perp = 0.7\omega_{p0}$ [113].

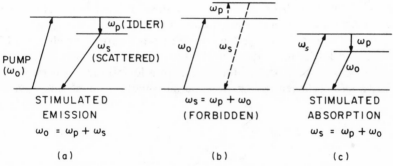

Figure 4.3 Energy-level diagrams of a three-wave FEL.

stimulated Raman absorption, and it is the Raman analog of the two-wave FEL accelerator.

The Stokes and anti-Stokes relationships are

$$\omega_{\text{Stokes}} = 2\gamma_{\|}^2 \left(k_0 v_{\|} - \frac{\omega_p}{\gamma} \right)$$

$$\omega_{\text{anti-Stokes}} = 2\gamma_{\|}^2 \left(k_0 v_{\|} + \frac{\omega_p}{\gamma} \right) \tag{4.40}$$

respectively, in the laboratory frame. The difference in frequency, $\Delta\omega_A$, between the two modes is just $2\gamma_{\|}^2(2\omega_p/\gamma)$, or

$$\frac{\Delta\omega_A}{\omega} \approx \frac{2\omega_p/\gamma}{k_0 \beta c} \approx 2\frac{\Delta\gamma_A}{\gamma} \tag{4.41}$$

If $v_{\|}$ is taken to be in resonance with k_0 and ω for the Stokes mode, then the signal will be amplified (Fig. 4.4a); however, if $v_{\|}$ or $\gamma_{\|}$ is reduced slightly, the electron will resonate with the anti-Stokes mode, and the signal wave will be absorbed (Fig. 4.4b). The wavevector of the space-charge wave is reversed, which corresponds to the signal wave accelerating the electron. Stimulated Raman absorption has been observed in an oscillator situation where the accelerator voltage was suddenly decreased [133]. Also, a Stokes wave propagating axially can be absorbed by an anti-Stokes wave propagating off axis at an angle

$$\theta_{0A} = \frac{1}{\gamma} \sqrt{\frac{2\omega_p}{\gamma k_0 c}} \tag{4.42}$$

To prevent interference, $\theta_{0A} > \lambda/2R$, the diffraction angle.

The parametric-amplifier version of the FEL becomes a parametric oscillator on the inclusion of optical feedback. Here we must make two changes. First, the waves grow spatially (convectively) as one moves down the undulator—parallel to the electron motion—in the laboratory frame; so one converts the temporal growth equations (4.23) and (4.27) to the laboratory frame, transforming the temporal rest-frame growth coefficient (4.28) using Eq. (4.1). Secondly, the optical resonator will have a loss coefficient ν_L for the optical wave (diffraction and mirror losses), and one might also justify a loss coefficient ν_i [e.g., Landau damping] for the plasma wave:

$$\frac{dE_s}{dz} = -i\Gamma E_i^* - \tfrac{1}{2}\nu_L E_s$$

$$\frac{dE_i^*}{dz} = +i\Gamma E_s - \tfrac{1}{2}\nu_i E_i^* \qquad (4.43)$$

BEAM FRAME

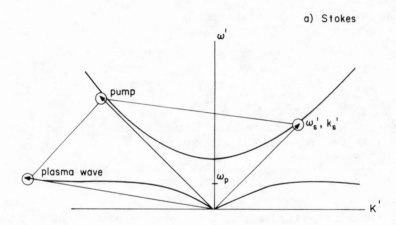

a) Stokes

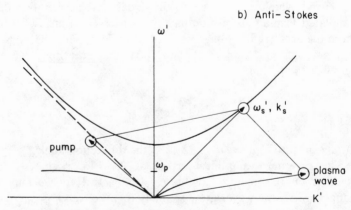

b) Anti-Stokes

Figure 4.4 Stokes diagrams (rest frame).

One can check this by setting ν_i, $\nu_L = 0$, and transforming the operator d/dz to the operator d/dt in the rest frame, in which case one recovers Eq. (4.29) with $\Gamma \sim g/2\gamma c$. The parametric-oscillator system will run in steady state when the losses just compensate for the gain, i.e., $d/dz = 0$. Setting the determinant of the coefficients of E_s and E_i equal to zero, the result for the threshold growth coefficient for oscillation is

$$\Gamma_{\text{thr}} = \sqrt{\nu_L \nu_i} \qquad (4.44)$$

This threshold has to do with the operation of the Raman FEL as an oscillator. However, if a more detailed theoretical treatment is followed to calculate the Raman growth rate [e.g., as in Section 4.3], another threshold is uncovered, connected with the appearance of the instability. A minimum value of the undulator field $B_\perp$ is necessary to sustain growth of the signal wave, and a yet larger value is needed to allow oscillation.

In an FEL oscillator, it is important to maintain the accelerator voltage constant within narrow limits, so that the frequency of the radiation stored in the optical resonator is very nearly in resonance with the electron beam energy. For the Raman FEL, this amounts to holding the variation of accelerator voltage within the tolerances shown in Fig. 4.1. Since g is bounded approximately by ω_p, for a submillimeter Raman FEL one finds that the accelerator voltage must be held constant to $\approx 1\%$ variation during many bounces of the radiation in the Fabry-Perot cavity—a difficult requirement for the pulse-line accelerators used to generate intense electron beams.

4.3 Traveling Wave Fluid Model

The model presented in Section 4.2 assumes that all waves are well-defined modes of the electron-beam system, and therefore it omits the phenomena that occur when the beat frequency does not excite such a normal mode (the plasma frequency). This means we can never recover the two-wave FEL, and worse yet, we cannot determine how space charge affects the performance of the two-wave FEL. A more general approach is needed. A very thorough analysis by Kroll and McMullin treats this situation in the context of a kinetic formulation [107]. A more simplified technique is used in this section, in which we use the cold-beam fluid equations; this is the method used by McDermott [133] and Marshall [8]. A more involved kinetic-theory method will be discussed in Chapter 6.

In Chapter 3 the electron energy loss appeared as radiation; however, when the beam is dense the energy loss of the particles is not necessarily transferred to the scattered wave, but can appear in the driven longitudinal oscillation. Rather than find the energy loss of the electrons, we calculate the temporal behavior of the scattered wave using the Laplace transform on a set of cold-fluid equations. The following theory is valid for beams having negligible thermal velocity spread, and we furthermore assume no transverse spatial dependence, no guiding magnetic field, and no pump depletion.

The process is analyzed in the beam frame, where the magnetostatic undulator field is an EM wave moving with phase velocity $\sim v_{\|} \sim c$. The amplitude of the pump electric wave is related to the transverse magnetic field by $E'_{\perp} = \gamma B_{\perp}$. The calculation is done in the rest frame, and for the remainder of this derivation the prime is dispensed with.

We take a "laser" ordering scheme in which the pump and scattered fields are first-order quantities and the quiver velocity induced at the beat frequency is second order. After Fourier transformation of the spatial variable, the coupled equations become

$$\left(\omega_p^2 + \frac{d^2}{dt^2} \right) \mathscr{E}(t) = -2\pi e n (1 + \alpha_{\perp}) \Omega_{\perp} v_s(t) e^{-i\omega_0 t} \tag{4.45}$$

$$\left[\left(c^2 k_s^2 + \omega_p^2 + \frac{d^2}{dt^2} \right) E_s(t) \right. $$
$$= 2\pi e n \left\{ \Omega_{\perp} u(t) e^{i\omega_0 t} - \left(\frac{i\alpha_{\perp}\Omega_{\perp}}{k_0} \right) \frac{d}{dt} \left[e^{i\omega_0 t} \left(\frac{n(t)}{n} \right) \right] \right\} \tag{4.46}$$

$$\frac{d}{dt} u(t) = -\frac{e}{m} \mathscr{E}(t) - \frac{1 + \alpha_{\perp}}{2} \Omega_{\perp} v_s(t) e^{-i\omega_0 t} \tag{4.47}$$

$$\frac{d}{dt} v_s(t) = -\frac{e}{m} E_s(t) \tag{4.48}$$

$$\frac{d}{dt} \left(\frac{n(t)}{n} \right) = i(k_0 + k_s) u(t) \tag{4.49}$$

where $\omega_p^2 = 4\pi n e^2/m\gamma$, ω_0 is the pump frequency, $\Omega_{\perp}/\omega_0 = a_\omega$, and $\alpha_{\perp} = [1 + (\Omega_{\perp}/\omega_0)^2]^{-1/2}$ represents the correction for the quiver-velocity dependence upon pump strength. A simple expression is taken for the pump magnetic field in the rest frame:

$$B_\omega(z, t) = \hat{y} \frac{B_{\perp}}{2} \left[e^{i(\omega_0 t + k_0 z)} + \text{c.c.} \right] \tag{4.50}$$

Equations (4.45) and (4.46) describe the generation of the longitudinal oscillation and scattered transverse wave in terms of nonlinear currents. Negligible beam space charge means that one can omit the first term on the left-hand side of (4.45). Equations (4.47) and (4.48) represent the force equations for the longitudinal and scattered fields, while (4.49) is the continuity equation for the longitudinal mode. These equations are valid for either circular or linear polarization of the pump wave.

As viewed in the beam frame, a strong, high-frequency velocity modulation is induced upon the electron beam as it flows through the pump field.

The quiver motion is only a few percent of c for Raman-type interactions, but in the short-wavelength high-energy FEL experiments it can be far larger. The ponderomotive force from the pump and scattered waves bunches the electrons [Eqs. (4.47) and (4.49)], strengthening the longitudinal beat wave (4.45), while the nonlinear current from the mixing of the quiver velocity and the density bunches drives the backscattered wave (4.46). By including the space-charge field $\mathscr{E}(t)$ and ω_p^2 the self-consistent beat potential is incorporated and we uncover collective effects in the appropriate limit.

The Laplace transform is applied to each equation, solving for the scattered field, after which the inverse Laplace transform is applied [8], yielding $E_s(t)$. After neglecting terms proportional to ω_i^2/ω_0^2, E_s can be written in terms of residues:

$$
\frac{E_s(t)}{E_s(0)e^{i\omega_s t}} = -\theta_p^2\theta_n \left\{ \frac{e^{-i\theta_1}}{\theta_1(\theta_1 - \theta_2)(\theta_1 - \theta_3)} \right.
$$
$$
\left. + \frac{e^{-i\theta_2}}{\theta_2(\theta_2 - \theta_1)(\theta_2 - \theta_3)} + \frac{e^{-i\theta_3}}{\theta_3(\theta_3 - \theta_1)(\theta_3 - \theta_2)} \right\}
$$

(4.51)

where θ_1, θ_2, and θ_3 are roots of the cubic

$$
-\theta_p^2\theta_n = \theta\left[\theta + (\theta_i + \theta_p)\right]\left[\theta + (\theta_i - \theta_p)\right]
$$

(4.52)

with $\theta_p = \omega_p T$, $\theta_i = \omega_i T$, $\theta_n = [\overline{\Omega_\perp^2}\gamma/2k_0 c]T$, in which T represents the time duration of the interaction as observed in the rest frame, $L/\gamma c\beta$. We define $\omega_i = \omega_0 - \omega_s$ and $\overline{\Omega}_\perp = \Omega_\perp[\alpha_\perp(1 + \alpha_\perp)/2]^{1/2}$.

The scattering rate is governed by the parameters θ_n, θ_p, and θ_i. Exponential growth can occur for $\theta_i \approx \theta_p$, as we have found in the preceding section. If we solve the last equation for $\theta_i \approx \theta_p$, we see that

$$
-\theta_p^2\theta_n = \theta^2\left[\theta + 2\theta_p\right]
$$

(4.53)

For $\theta_n \ll \theta_p$, the exponential growth parameter is $(\theta_p\theta_n/2)^{1/2}$, and if this number is > 1, then the electric field of the scattered wave is given by

$$
E_s(T) = E_s(0)\cosh\left(\frac{\theta_p\theta_n}{2}\right)^{1/2} e^{i\omega_s T}
$$

(4.54)

provided $\omega_i = \omega_p$, that is, when the beat wave has a distinct identity. An example is shown in Fig. 4.5. Here the gain maximizes at $\omega_i = \omega_p$, and at $\omega_i = -\omega_p$ we have the case of anti-Stokes, or stimulated Raman, absorption. For $\omega_i \neq \omega_p$, the gain is very much smaller, and represents the two-wave beat process which still occurs; note this is quite small but that Eq. (3.30)—the two-wave gain result—would (erroneously) predict an extremely high gain.

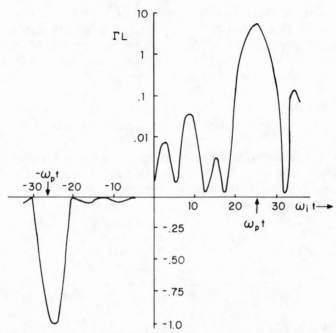

Figure 4.5 Dependence of Raman FEL growth rate on beat frequency of idler [133]; $B_\perp = .5$ kG, filling factor = 1, $L = 50$ cm. © 1980 Addison-Wesley.

For $\theta_n \gg \theta_p$, the result is divided into two regimes. If $\theta_p^2 \theta_n \gg 1$, exponential growth for $\theta_i \approx \theta_p$ is still dominant with exponent $\frac{1}{2}\sqrt{3}\,(\theta_p^2\theta_n)^{1/3}$; this is called the strong-pump regime or oscillating two-stream instability. Here the strength of the pump and the ponderomotive potential dominate the collective effects, and the plasma wave is subordinated. Figure 4.6 [2] shows how the Raman and strong-pump growth rates blend into each other. The efficiency increases in the strong-pump regime, being larger by a factor $\approx \frac{1}{2}(\Omega_\perp^2 \gamma/\omega_p k_0 c)^{1/3}$ than the Raman efficiency. On the other hand, if $\theta_p^2 \theta_n \ll 1$, we have the tenuous-beam limit, the exponential growth is unimportant, and the gain mechanism is dominated by interference effects. Since two of the roots of Eq. (4.52) lie near $-\theta_i$, the normalized beat frequency, two of the residues in Eq. (4.51) may be combined approximately into the form of a derivative. After this derivative is combined with the remaining residue, the power amplification or gain—defined as

$$G \equiv \frac{E_s^*(t)E_s(t) - E_s^*(0)E_s(0)}{E_s^*(0)E_s(0)} \qquad (4.55)$$

—can be shown to satisfy, for small θ_p,

$$G = -4\theta_p^2\theta_n\frac{\partial}{\partial\theta_i}\left(\frac{\sin\frac{1}{2}\theta_i}{\theta_i}\right)^2 \qquad (4.56)$$

which has already been derived in Chapter 3 (shown in Fig. 3.8) using a completely different formulation. This result predicts that for a given undulator length the gain will be maximized by a specific mismatch between the pump and signal frequencies—in particular, for $\omega_0 - \omega_s = 2.6/T$, in which case $G = G_N$, i.e.,

$$G_N = 0.27\theta_p^2\theta_n \qquad (4.57)$$

The normalized gain at $\theta_i = 2.6$ for various values of θ_n (pump strength) is obtained numerically from Eqs. (4.51) and (4.52) and is plotted in Fig. 4.7 as a function of θ_p [133]. Curve 3 corresponds roughly to the first Stanford FEL for its value of θ_n, taking $\theta_p = 0.1$. For a weak pump (curve 1), the gain is reduced below G_N by the inclusion of space charge, whereas for a strong pump (curve 3) the gain can be enhanced by the inclusion of space charge. In finding first-order corrections to G_N by retaining space charge, Louisell [124] calculated a weak-pump example.

The relationship between the collective FEL and the two-wave FEL is contained in this analysis. The Raman FEL is operated in the high-density, moderate-pump regime; here a three-wave decay interaction with a plasma-wave idler will cause exponential growth of the scattered wave. The two-wave FEL is characterized by low plasma frequency and high pump amplitude. The gain of the latter, due to interference effects, may be enhanced through the space-charge interaction.

We shall now compare the efficiencies of the two- and three-wave FELs (an interesting exercise, but not a practical one, as the two FELs operate in different domains of the spectrum). In the two-wave case, electrons are injected above resonance and can at most be decelerated near resonance

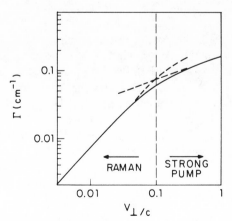

Figure 4.6 Raman-FEL spatial growth-rate variation between Raman and strong-pump regimes [2]. © 1982 CRC.

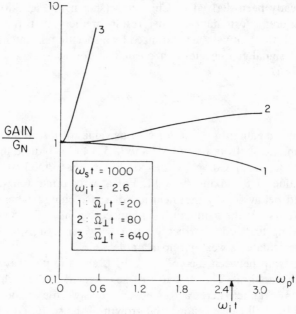

Figure 4.7 Effect of space charge on two-wave FEL gain [133]. © 1980 Academic Press.

($\omega_i = 0$); since $\omega_s = 2\pi\gamma\beta c/l_0$ in the beam frame, ω_s can change at most by

$$\Delta\omega_s = \frac{2\pi\gamma c}{l_0}(\Delta\beta)' \approx \omega_i = \frac{2.6}{T} = \frac{2.6\gamma\beta c}{L} \qquad (4.58)$$

We can solve the above for $\Delta\beta'$, and use

$$\eta_{2\text{-w}} = \left(\frac{\Delta\gamma}{\gamma}\right)_{2\text{-w}} = \Delta\beta' = \frac{2.6}{2\pi N} \sim \frac{1}{2N} \qquad (4.59)$$

for the two-wave FEL (the same result as obtained in Chapter 3). The efficiency is proportional to $\Delta\gamma/\gamma$, which is also proportional to a frequency shift $\Delta\omega/\omega$. In the two-wave FEL, this "frequency shift" is of order $2.6/T$ for this purpose. For the three-wave FEL, the frequency shift is of order g [See eq. (4.35)], which can be no larger than ω_p. Thus the ratio of the Raman FEL efficiency to the two-wave FEL efficiency is of order $\omega_p T$ [more accurately, it is $2/\pi\omega_p(L/\gamma c)$]. The collective FEL is more efficient in the regime where it can compete with the two-wave FEL.

4.4 Effects Caused by Electron Velocity Spread

Warm-beam effects can be calculated by recasting the equations (4.45)–(4.59) in a kinetic-theory formulation (see Chapter 6), but the solution is difficult except in a few limiting cases. When the beam velocity distribution is Lorentzian in the rest frame, viz.

$$f_e(v_z') = \frac{v_T'}{\pi} \frac{1}{v_z'^2 + v_T'^2} \tag{4.60}$$

the calculation is comparatively simple; the wave growth is reduced in all regimes of FEL operation when $v_T'/c = (\delta\gamma/\gamma)_{\parallel} \gtrsim 1/N$. However, the Lorentzian is taken only for convenience, does not represent a thermal distribution, and contains an excess of high-velocity particles.

When the electron beam is warm, i.e. when $(\delta\gamma/\gamma)_{\parallel} > 1/N$, a kinetic formulation of the problem outlined by Eqs. (4.45)–(4.49) is needed to calculate the exponential gain coefficient. The calculation assumes a warm-beam distribution $f_e(v_z') = (1/\sqrt{2\pi}\,v_T')e^{-(v'/v_T')^2}$. A simple result for the exponential growth rate was obtained by Hasegawa [95, 96] for the case $v_\perp' = eB_\perp/k_0 mc < v_T'$:

$$g = 0.2\left(\frac{\omega_p^2}{\gamma k_0 c}\right)\left(\frac{v_\perp'}{v_T'}\right)^2 \tag{4.61}$$

In this example, $v_T' \gtrsim \omega_p \gamma l_0/4\pi$ and the scattering is due to electrons having velocity nearly equal to the phase velocity of the beat wave, $\omega_i/(k_0 + k_s)$. The growth rate of the radiation is low because of the small number of electrons participating in the scattering, and saturation is quickly reached.

The relationship between the cold-beam and the warm-beam Raman FEL is shown in Fig. 4.8. In case (a), there are no electrons for the beat wave to interact with; the distribution is cold for practical purposes even though there is some width. In case (b), the phase velocity of the space-charge wave falls within the envelope of the electron-beam velocity spread. Then there are electrons which are moving at nearly the speed of the wave, and—depending on the slope of $f_e(v)$—the wave interacts with these electrons in an important way. For $f_e'(v) < 0$, more electrons receive energy from the wave than release energy to the wave, and the situation is referred to as Landau damping. The amplitude of the space-charge wave would, if energy were not supplied by the FEL interaction, damp away in a few cycles. As the damping becomes of order $\omega_p(L/\gamma c\beta)$ periods, the identity of the "third" wave is blurred and one makes a transition to the case of Eq. (4.61).

We now endeavor to establish this criterion more quantitatively. The transformation between the electron velocity u_z' in the rest frame and u_z in the laboratory frame is given by

$$u_z = \frac{u_z' + v}{1 + vu_z'/c^2} \tag{4.62}$$

A beam will be characterized by a range of electron velocities, $\delta u_z'$ or δu_z in the respective frames:

$$\delta u_z = \frac{\delta u_z'}{\left(1 + vu_z'/c^2\right)} - \frac{u_z' + v}{\left(1 + vuz'/c^2\right)^2}\left(\frac{v}{c^2}\right)\delta u_z' \qquad (4.63)$$

or

$$\delta u_z \approx \delta u_z'\left(1 - \frac{v^2}{c^2}\right)$$

Therefore

$$\delta u_z' = \gamma^2 \delta u_z \qquad (4.64)$$

In the laboratory frame, one finds by differentiation that

$$\left(\frac{\delta\gamma}{\gamma}\right)_{\shortparallel} = \gamma^2 \frac{\delta u_z}{c} \qquad (4.65)$$

and therefore

$$\left(\frac{\delta\gamma}{\gamma}\right)_{\shortparallel} = \frac{\delta u_z'}{c} \qquad (4.66)$$

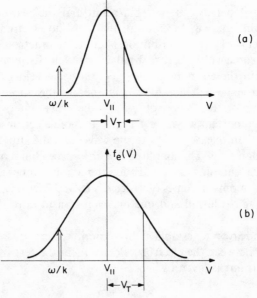

Figure 4.8 "Hot" (b) and "cold" (a) electron velocity distribution functions and the space-charge wave phase velocity.

In order that the space charge wave be relatively undamped, one requires that its wavelength ($l_0/2\gamma$ in the rest frame) be larger than the beam Debye distance, defined as

$$\lambda_D' = \frac{v_T'}{\omega_p} = \frac{\delta u_z'}{\omega_p} \tag{4.67}$$

in the rest frame. Equation (4.67) applies for a thermal (Gaussian) beam velocity distribution of a nonrelativistic electron system only. The nature of the Landau-damping process is such that if the space-charge wavelength is only twice λ_D', then the Landau damping constant is about ten ω_p cycles of the wave—a number which we have previously chosen to establish the collective regime in a sufficiently dense electron beam passing down an undulator of finite length [Eq. (2.60)]. Therefore, finite parallel velocity spread compromises the Raman interaction when

$$\frac{l_0}{2\gamma} \approx 2\frac{v_T'}{\omega_p} = 2\frac{\delta u_z'}{\omega_p} \tag{4.68}$$

and so we require the following inequality:

$$\left(\frac{\delta\gamma}{\gamma}\right)_{\|} < \left(\frac{l_0}{2\gamma}\right)\left(\frac{\omega_p}{2c}\right) \tag{4.69}$$

Note c/ω_p is the electromagnetic skin depth. If we take the liberty of choosing $\omega_p T \approx 4$ [from Eq. (2.60)], there is a memorable result:

$$\left(\frac{\delta\gamma}{\gamma}\right)_{\|} < \frac{1}{N} \tag{4.70}$$

or, thermal effects on the beam become important when the normalized momentum spread becomes of order N^{-1}. One must take N rather large (≈ 50) to develop FEL coherence and gain, so this implies that—in addition to high electron density—we require the beam to be cold, at least within the limit mentioned above. Replacing the quantity $l_0/2\gamma$ in Eq. (4.69) by the FEL wavelength, the condition on the wavelength for Raman FEL operation becomes $\lambda_s > (2/\gamma)(c/\omega_p)(\delta\gamma/\gamma)_{\|}$; since $c/\omega_p \sim 1$ cm, typically the Raman FEL must operate at a wavelength $\gtrsim 2/\gamma N \sim 10$–$100$ μm.

   To summarize, if $(\delta\gamma/\gamma)_{\|} < N^{-1}$, one can use the Raman formula for exponential signal gain. If $(\delta\gamma/\gamma)_{\|} \gtrsim N^{-1}$, then exponential growth still obtains, albeit with diminished vigor [given by Eq. (4.61)]. At some point, the declining exponential growth may become comparable with the two-wave gain, although similar considerations apply to the effect of beam quality on the two-wave FEL.

In the case of the two-wave FEL, we note that the gain is proportional to the derivative of the spontaneous radiation spectrum. The natural fractional width of that spectrum is $\sim 1/N$. If the spread of electron energies in the beam satisfies $(\delta\gamma/\gamma)_{\parallel} < 1/N$, no additional broadening will occur, and the gain will be as calculated in Chapter 3. If $(\delta\gamma/\gamma)_{\parallel} > 1/N$, the spontaneous spectrum will be broadened and the gain will decrease [see Section 5.5 under "Effect of Inhomogeneous Broadening on Gain"]. If the two-wave FEL operates on harmonic number p, then $(\delta\gamma/\gamma)_{\parallel} \ll 1/pN$ is required A quantitative study showing how the two-wave FEL gain decreases with increasing electron momentum spread has been made by Colson and Ride [47].

4.5 The Generalized Pendulum Equation

One might hope that the simple procedure for calculating FEL behavior we developed in Section 3.3 could be generalized so that it could treat cases where the beam density is high. In fact, a generalization of the pendulum equation has been found [183], and it is adaptable to the variable-parameter undulator problem. This section therefore represents a mathematically more precise summary of that problem, which is given in enough detail to permit use in actual calculations.

The vector potentials for a linearly polarized undulator and radiation fields may be represented by

$$A_\omega(z) = \hat{\imath} A_\omega(z)\cos\left\{ \int_0^z k_0(z')\, dz'\right\}$$

$$A_s(z) = \hat{\imath} A_s(z)\sin\left\{ \frac{\omega z}{c} - \omega t + \phi(z)\right\} \qquad (4.71)$$

where $A_\omega(z)$, $A_s(z)$, ϕ, and k_0 are slowly varying functions of z. The ponderomotive potential again depends on $\sin\psi$, and the phase difference between the electron and the ponderomotive wave is

$$\psi(z,\psi_0) = \int_0^z\left[k_0(z') + \frac{\omega}{c} - \frac{\omega}{v_z}(z',\psi_0)\right] dz' + \phi(z) + \psi_0 \quad (4.72)$$

Here ψ_0 is the initial phase at the entrance of the undulator, and the axial electron velocity is

$$v_z = \omega\left[\frac{\omega}{c} + k_0 - \frac{\partial\psi}{\partial z} + \frac{d\phi}{dz}\right]^{-1} \qquad (4.73)$$

Using the conservation of canonical momentum [see Eq. (6.6)], there results the *generalized pendulum equation*:

$$\frac{d^2\psi}{dz^2} = \frac{d^2\phi}{dz^2} + \frac{dk_0}{dz} - \frac{\omega}{4c}\left(\frac{e}{\gamma mc^2}\right)^2 \frac{\partial A_\omega^2}{\partial z} - \frac{\omega}{c}\left(\frac{e}{\gamma mc^2}\right)^2 k_0 A_\omega A_s \cos\psi$$

$$+ 2\frac{\omega_p^2}{c^2\gamma_\parallel^2}\left[\langle\cos\psi\rangle_{\psi_0}\sin\psi - \langle\sin\psi\rangle_{\psi_0}\cos\psi\right] \qquad (4.74)$$

The operation of averaging over initial phases is defined by

$$\langle(\cdots)\rangle_{\psi_0} \equiv \frac{1}{2\pi}\int_0^{2\pi}(\cdots)\,d\psi_0$$

The first, second, and third terms of Eq. (4.74) represent respectively the effect on ψ due to the variation of the phase of the radiation field, the undulator period, and the undulator amplitude. The effect of the ponderomotive term is given in the fourth term, while the last term supplies the effect of the space-charge wave. Comparing the last two terms, the condition on the beam density appropriate to neglecting the space charge is seen to be

$$n \ll \frac{k_0^2\gamma^3 A_\omega A_s}{4\pi mc^2} \qquad (4.75)$$

The radiation field, characterized by its amplitude and phase, is obtained by solving the field eqs. 3.31:

$$\left(\frac{\omega}{c} - K\right)A_s = \frac{\omega_{p0}^2}{c^2}\left(\frac{c}{2\omega}\right)A_\omega\left\langle\frac{\sin\psi}{\gamma}\right\rangle_{\psi_0}$$

$$K^{1/2}\frac{d}{dz}\left(A_s k^{1/2}\right) = \frac{1}{2}\frac{\omega_{p0}^2}{c^2}A_\omega\left\langle\frac{\cos\psi}{\gamma}\right\rangle_{\psi_0} \qquad (4.76)$$

where

$$K = \left(\frac{d\phi}{dz} + \frac{\omega}{c}\right)$$

The electron energy changes according to

$$\frac{\partial}{\partial z}\left[\gamma(z,\psi_0)mc^2\right] = -\frac{\omega/c}{2\gamma}\left(\frac{e^2}{m^2c^4}\right)A_\omega A_s \cos\psi \qquad (4.77)$$

The electron with constant phase is the "resonant" particle, and it has resonant γ_r defined by

$$\gamma_r = \gamma_\perp \gamma_{zr} \qquad (4.78)$$

where

$$\gamma_{zr} = \sqrt{2} \left\{ \frac{K}{k_0 + d\phi/dz} \right\}^{1/2}, \qquad \gamma_\perp \equiv \left(1 + a_\omega^2\right)^{1/2}$$

To maintain an electron at resonance, one can prescribe either $k_0(z)$ or $A_\omega(z)$ by substitution into

$$0 = \frac{dk_0}{dz} + \frac{d^2\phi}{dz^2} - \frac{\omega/c}{2\gamma_r^2} \left(\frac{e^2}{m^2 c^4}\right) \left[\frac{\partial}{\partial z}\left(\frac{A_\omega^2}{2}\right) + 2k_0 A_\omega A_s \cos\psi_r\right] \tag{4.79}$$

The fraction of electrons trapped depends on the amplitude of the ponderomotive potential and the axial electron velocity spread δu_z. The condition (see eq. 3.18)

$$\left(\frac{\delta\gamma}{\gamma}\right)_{\parallel} < 2^{3/2} \left(\frac{\gamma_{\parallel}}{\gamma}\right)\left(\frac{e}{mc^2}\right)\sqrt{A_\omega A_s} \tag{4.80}$$

defines the requirements on the axial beam momentum spread so that the electrons will interact with the ponderomotive wave as if they were cold. It is interesting to observe that a beam with $(\delta\gamma/\gamma)_{\parallel}$ too large to sustain

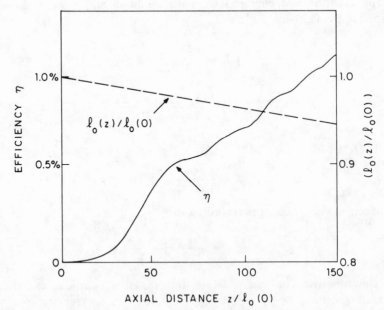

Figure 4.9 An Efficiency-enhancement calculation by Tang [190]; variable-period undulator. © 1981 AIP.

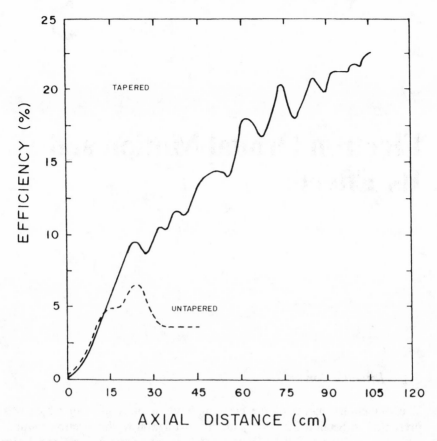

Figure 4.10 Comparison of a Raman FEL with and without tapering of undulator period [191].

effective small signal gain, may still be acceptable here providing the incoming signal amplitude is sufficiently large.

An example of how FEL efficiency can be enhanced by "programming" the undulator is shown in Fig. 4.9. In this example, if the undulator period were not tapered, the FEL gain would be zero, as the ponderomotive wave was chosen to be exactly resonant with the electrons at the entry to the undulator [190]. The formulation of the nonuniform undulator problem above shows that the efficiency enhancement methods developed for the two-wave FEL will apply as well for the Raman FEL. Efficiency enhancements as large as ≈ 20% can be obtained for the tapered-period undulator in conjunction with the three-wave FEL (Fig. 4.10 [191]).

5

Electron Orbital Motion and Its Effects

5.1 Introduction

The elementary model of the electron orbit in the undulator which was presented in Section 2.2 is only an approximation to the electron motion, which is basic to the FEL physics. Also, there are several types of undulator, with rather important differences in field description. Theorists have taken liberties with the modeling of electron motion in these undulators. When we realize that the entire theory of the FEL must rest ultimately on a kinetic equation insofar as the electrons travel in nonidentical orbits, the situation becomes rather serious and complicated. One must grapple with orbit problems for three main reasons: to understand where approximations are justified; to calculate FEL physics correctly in situations where theory requires the orbit; to account for sources of inhomogeneous line broadening.

Taking a Socratic approach, are the electron orbits in a helically wound undulator helical, or only approximately so? If departure from helicity is bad, how can one construct a better undulator? If the orbits are not helical, how can this adversely affect the FEL—or open new possibilities? Are the orbits stable, or do small departures from helicity (e.g., betatron oscillations) grow and peel electrons away from the beam? If there are necessary limitations on orbital behavior, what corresponding limitations apply to the FEL gain, efficiency, etc.? There are many ramifications.

We begin this chapter with a brief example of electron-beam equilibrium, reminding the reader that in general this is actually the point of departure for orbital calculations in the new or perturbed equilibrium that is set up as the electrons stream into the undulator. In Sections 5.3 and 5.4 we study the magnetic fields and orbits in certain undulators, attempting to outline what has been done and where complications arise. The next section contains a summary of the sources of inhomogeneous line broadening in the FEL; these effects arise because not all orbits are identical and this causes the axial electron motion to be nonuniform.

5.2 An Electron-Beam Equilibrium

As electrons stream parallel to each other in an electron beam, they experience forces from external fields (a guiding or focusing magnetic field, as a rule) and from the fields generated by the collection of electrons (an electric field from the space charge and a magnetic field from the current). At highly relativistic energy, the two self-forces almost balance: the space-charge force is radially outward, whereas the current produces an inward "pinching" force. For a beam with uniform electron density in cylindrical geometry,

$$\left. \begin{array}{l} E_r = 2\pi enr \\ B_\theta = 2\pi enr\beta \end{array} \right\} \quad r < r_b \tag{5.1}$$

and hence the total force from the internal fields is

$$e\left[E_r + \frac{v_z}{c} \times B_\theta \right] = 2\pi e^2 nr(1 - \beta^2) \tag{5.2}$$

or, since $I_b = \pi r_b^2 ne\beta c$ is the beam current,

$$= \frac{2I_b er}{r_b^2 \beta c \gamma^2}$$

The beam current can be expressed in terms of Budker's parameter $\nu = \pi r_b^2 n(e^2/mc^2)$ and the Alfven current $I_A = 17{,}000\beta\gamma$ A, whereupon $I_b = 17{,}000\nu\beta$ A. For even rather high-current beams ($> 10^3$ A/cm^2), it is necessary that $\nu/\gamma \ll 1$ to insure a cold beam unless the beam is very thin and a very strong magnetic field is used.

There are many equilibria for electron beams propagating with certain simple configurations along or into a guiding field [1]. In FEL research we are interested in only those equilibria for which the electrons move at nearly constant relativistic speed along the axis of the system. In what follows, we discuss a simple example which is representative of the type of low-density, uniform electron beam having cylindrical symmetry which is used in FEL applications.

In this section we shall disregard the undulator. The system conserves energy:

$$(\gamma - 1)mc^2 + ev(r) = -eV_k \tag{5.3}$$

where V_k is the cathode potential and $V(r)$ is the electrostatic potential in the beam. There is symmetry about the z-axis, and so the canonical angular momentum P_θ is a constant of the motion:

$$P_\theta(r) = \gamma(r)mrv_\theta(r) - \frac{er}{c}A_\theta(r)$$

$$= -\frac{er_k}{c}A_{\theta k}(r_k) \tag{5.4}$$

where k refers to the cathode, $A_\theta \approx rB_0/2$, and v_θ is the azimuthal component of velocity. If $B_0(r_k) = 0$ then $P_\theta = 0$; on the other hand, the cathode may be immersed in the magnetic field, in which case $P_{\theta k} \alpha r^2 B_{0k}$. A beam can also be emitted as a cylindrical shell which flows along a flux surface, whereupon $P_\theta = $ constant.

Certain equilibria may be approximated by a rigid rotor azimuthal motion of the beam about the symmetry axis. We shall assume $\beta_z(r) \approx$ constant, and suppose that the guiding field is also constant across—as well as along—the beam (this means that the rigid-rotor motion is not too fast). Gentle rigid-rotor motion also implies the centrifugal force is not large, and it is therefore reasonable to take the charge density nearly constant across the beam. The radial force-balance equation is

$$0 = \frac{\gamma mv_\theta^2}{r} + \frac{2\pi e^2 nr}{\gamma^2} - \frac{v_\theta B_0 e}{c} \tag{5.5}$$

The first term is centripetal, the second term is due to the combined effect of the internal fields [Eq. (5.2)], and the third is the $v_\theta \times B_0$ component of the Lorentz force. Taking $\omega_{p0}^2 = 4\pi ne^2/m$, one finds a rigid-rotor motion is indeed possible, that is, $\omega_\theta = v_\theta/r$ is a constant:

$$\omega_\theta^\pm = \frac{\Omega_0}{2\gamma}\left[1 \pm \left(1 - \frac{2\omega_{p0}^2}{\gamma\Omega_0^2}\right)^{1/2}\right] \tag{5.6}$$

This places a limit on the electron density:

$$2\omega_{p0}^2/\gamma \leq \Omega_0^2 \tag{5.7}$$

The larger ω_θ^+, corresponds to the fast gyration of the electron about the lines of magnetic force (in the limit $\omega_{p0} \to 0$, $\omega_\theta^+ \to \Omega_0/\gamma$). The smaller root is

$$\omega_\theta^- \approx \frac{\omega_{p0}^2}{2\gamma^2\Omega_0} \approx \frac{cE_r}{r\gamma^2 B_0} \tag{5.8}$$

and is an $E_r \times B_0$ azimuthal drift in the space-charge electric and guiding magnetic fields. The two roots merge if

$$\Omega_0^2 = \frac{2\omega_{p0}^2}{\gamma} \quad \Rightarrow \quad \omega_\theta^\pm = \frac{\Omega_0}{2\gamma} \tag{5.9}$$

and this simple rotation is referred to as *Brillouin flow*.

Typical operating conditions for a high-current-density beam would be $\omega_{p0}/2\pi \approx 3$ GHz, $\gamma \approx 3$, $\Omega_0/2\pi \approx 30$ GHz, with $r_b \approx \frac{1}{4}$ cm: then $v_\theta/c \approx 0.2\%$ at $r = r_b$ due to the space-charge rotation. Figure 5.1 shows how the electron orbit in an undulator is modified by the drift motion. The large cycloids have to do with the $E_r \times B_0$ motion, whereas the small "scallops" are caused by the undulator in this simulation. Considerable insight into electron motion in the beam, when an undulator is present, can be obtained by numerical calculation of the orbits: in Fig. 5.1 the undulator fields were represented by Eq. (5.13), while the self electric and magnetic fields used were calculated from the global properties of the beam, so the orbit is that of a "test particle" [89].

Does rigid-rotor equilibrium obtain for the simple case where the beam follows uniform magnetic-field lines from a cathode? In this example, it can

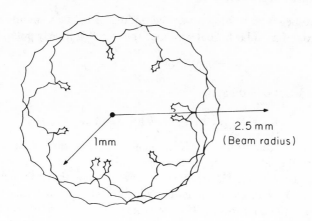

(x - y PLANE)

Figure 5.1 Orbit of a test electron in beam self-fields and field of an undulator (transverse plane) [89].

be shown that the rigid-rotor and constant-axial-velocity equilibria are inconsistent [159]. Optimization of the beam parallel-velocity profile would in practice require extensive numerical analysis, directed toward a special diode and cathode configuration. The exception is when the source is in a field-free region.

A comprehensive treatment of the equilibrium and stability of relativistic electron beams is contained in a monograph by Davidson [1], as well as in texts by Miller [9] and Lawson [7].

5.3. *The Bifilar Helical Undulator*

One must bear in mind that there is necessarily a difference between the actual motion of electrons in an FEL and that which is used as a theoretical model. Hopefully the discrepancies will not be important if the undulator is well designed, or the electron beam is the right size, at the right location. For a number of reasons having to do with experimental constraints, this is not always possible. For example, a popular theoretical model of a helical undulator field is a reasonable approximation on the axis of the solenoid, for sufficiently small electron orbit radius. But we have already seen that diffraction considerations suggest the use of a wider beam, and of course the FEL gain will improve if we turn up the undulator field (electron orbit radius). Optimization of FEL performance—or for that matter, any performance at all—may require a better model.

Magnetic-Field Representation

Suppose we begin by writing down the undulator field commonly used very near the axis of a bifilar helical winding; this is a "circularly polarized" field

$$\boldsymbol{B}_\omega = B_\perp \left(\hat{\boldsymbol{\imath}} \cos k_0 z + \hat{\boldsymbol{\jmath}} \sin k_0 z \right) \tag{5.10}$$

or, in cylindrical coordinates,

$$\boldsymbol{B}_\omega = B_\perp \left[\hat{\boldsymbol{r}} \cos(k_0 z - \theta) + \hat{\boldsymbol{\theta}} \sin(k_0 z - \theta) \right] \tag{5.11}$$

In such a field the transverse canonical momenta are constants of the motion (in the absence of a guiding field), and the resulting electron trajectories are helices. However, it is easy to show that this field fails to satisfy $\nabla \times \boldsymbol{B}_\omega = 0$, and so it is unrealizable. This implies that the actual electron orbits depart from helices to some degree. There is also one other problem: Real undulators are made of wires or windings which do not carry simple sinusoidal surface currents. Accordingly, the actual field will have higher-order harmonic components than k_0, and so also will the electron motion. This situation can be improved by careful engineering.

The field from a bifilar helical winding, excited by a sinusoidal surface current, has been calculated [34, 105]. The components of field are

$$B_r = B_\perp \left[2I_1'(k_0 r)\right] \sin(\theta - k_0 z)$$

$$B_\theta = B_\perp \frac{2I_1(k_0 r)}{k_0 r} \cos(\theta - k_0 z)$$

$$B_z = -B_\perp \left[2I_1(k_0 r)\right] \cos(\theta - k_0 z) \tag{5.12}$$

where the I-functions are Bessel functions of imaginary argument. These fields satisfy both $\nabla \times \boldsymbol{B}_\omega = 0 = \nabla \cdot \boldsymbol{B}_\omega$. For $k_0 r \lesssim 0.8$ these can be approximated by

$$B_r = B_\perp \left(1 + \tfrac{3}{8}k_0^2 r^2\right) \sin(\theta - k_0 z)$$

$$B_\theta = B_\perp \left(1 + \tfrac{1}{8}k_0^2 r^2\right) \cos(\theta - k_0 z)$$

$$B_z = -B_\perp (k_0 r)\left(1 + \tfrac{1}{8}k_0^2 r^2\right) \cos(\theta - k_0 z) \tag{5.13}$$

Note that $B_z \to 0$ as $r \to 0$, whereas $B_{r,\theta}$ are finite. The transverse field amplitude on the axis of the undulator is related to the current in the windings, I_w, by the following approximate result:

$$B_\perp(0) \approx \frac{4\pi^2}{5}\left(\frac{I_w}{l_0}\right)\left(\frac{a_0}{l_0}\right)^{1/2} e^{-2\pi a_0/l_0} \tag{5.14}$$

where $B_\perp$ is in gauss, I_w in amperes, and l_0 and a_0 (the helical-wire radius) in centimeters. The current can become prohibitively large if $2\pi a_0/l_0 > 1$. An exact treatment of this problem, including the effect of the windings, has been given by Park [150].

Orbits

If we represent the undulator field by the idealized form (5.10), that is, using a vector potential

$$A_\omega(z) = -\frac{B_\perp}{k_0}\left[\hat{\imath}\cos k_0 z + \hat{\jmath}\sin k_0 z\right] \tag{5.15}$$

then the transverse canonial momenta,

$$P_\perp = m\gamma v_\perp - \frac{|e|A_\omega}{c} \tag{5.16}$$

are constants. The electrons execute helical orbits with energy-dependent amplitude r_0 and constant axial velocity according to

$$k_0 r_0 = \frac{v_\perp}{v_{\parallel}} = \frac{|e|B_\perp}{\gamma mck_0 v_{\parallel}}$$

$$v = \hat{z}v_{\parallel} + \frac{|e|B_\perp}{\gamma mk_0 c}\left[\hat{\imath}\cos(k_0 v_{\parallel}t + \phi) + \hat{\jmath}\sin(k_0 v_{\parallel}t + \phi)\right]$$

$$\gamma = \text{constant} \tag{5.17}$$

An unpleasant but frequently necessary complication is to require in addition a constant guiding magnetic field. A detailed study of electron motion in the combined axial and helical undulator fields has been done by Friedland [77] for the idealized case; it has been revised by Freund et al. [73] for the realizable undulator. The quiver velocity of the electron is

$$v_\perp = \frac{2\Omega_\perp v_{\parallel}\left[I_1(\lambda)/\lambda\right]}{\Omega_0 - \gamma k_0 v_{\parallel} \pm 2\Omega_\perp I_1(\lambda)} \tag{5.18}$$

which simplifies [see Eq. (2.8)] when $\lambda = k_0 r_0$ is small. Two classes of stable orbit are found: Group I ($\Omega_0 < \gamma k_0 v_{\parallel}$) and Group II ($\Omega_0 > \gamma k_0 v_{\parallel}$), which are shown in Fig. 5.2 for $\gamma = 3.5$ and $\Omega_\perp/\gamma k_0 c = 0.5$ ($\beta_0 = \Omega_0/\gamma k_0 c$). The portions of the curves labeled "unstable" correspond to orbits where a betatron-oscillation perturbation becomes an exponential instability.

Friedland also studied various stratagems for injecting electrons into the undulator from a region of uniform magnetic field. In general one can

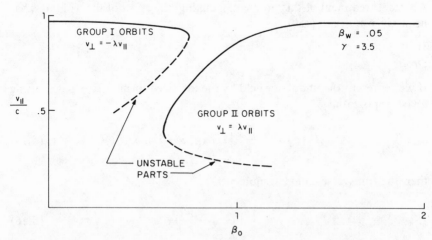

Figure 5.2 Orbit parameters of electrons in a guiding field, showing magnetoresonance effect [73]. © 1983 IEEE.

obtain a stable, nearly helical orbit in the undulator provided the parameters of the undulator are varied slowly throughout the transition region. This might require that the undulator field amplitude increase gradually over several periods of electron gyromotion and several periods of the undulator. The point of orbital instability is avoided, for constant guiding field, provided [18] $B_\perp/B_0 \lesssim [(\gamma^2 - 1)^{1/3}(k_0 c/\Omega)^{2/3} - 1]^{3/2}$.

In Fig. 5.3 is shown a numerical simulation for the electron orbit—displayed in the transverse plane—that results when the electron is launched into an undulator field which is tapered over several periods [89]. The electron spirals out as it picks up transverse motion, and settles in a nearly stable orbit in the uniform section of the undulator (the heavy black line indicates near-overlapping of many spirals). The orbital scale length is $\sim k_0 r_0$ [Eq. (5.17)].

Solution for the electron orbit has been made by Freund [71, 72] and by Uhm [192] for the idealized case including a guiding magnetic field:

$$\boldsymbol{B} = B_0 \hat{z} + B_\perp [\hat{i}\cos k_0 z + \hat{j}\sin k_0 z] \tag{5.19}$$

They found a result corresponding to nearly uniform axial velocity, where the electron motion predominantly follows the helical undulator but also contains a small component of Larmor precession in the axial field (Fig. 5.4). The "constants" of motion are given by $p_\parallel = \gamma m v_\parallel$ together with P_x

Figure 5.3 Orbit of a test electron moving into a tapered undulator field (transverse plane) [89]; motion is axial at undulator entrance.

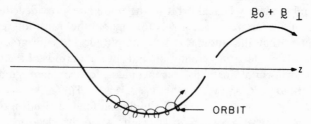

Figure 5.4 Representation of electron spiraling around a field line, in combined undulator and gyromotion.

and P_y, the components of $P_\perp$. If $v_\perp$ is not too large, then

$$v_x(t) = v_\perp \cos k_0 z + v_{\perp 0} \cos\left(\frac{\Omega_0}{\gamma} t + \phi\right)$$

$$v_y(t) = v_\perp \sin k_0 z + v_{\perp 0} \sin\left(\frac{\Omega_0}{\gamma} t + \phi\right)$$

$$v_z(t) = v_\parallel - \Delta v_z(t) \tag{5.20}$$

where $v_\perp = \Omega_\perp V_\parallel / (\Omega_0 - \gamma k_0 v_\parallel)$, $v_{\perp 0}$ and ϕ are constants which describe the Larmor gyromotion in the axial guide field, and

$$\Delta v_z(t) \approx \frac{\Omega_\perp v_{\perp 0}}{\Omega_0 - \gamma k_0 v_\parallel} \cos\left(k_0 z - \frac{\Omega_0 t}{\gamma} - \phi\right) \tag{5.21}$$

This solution is not valid as $\Omega_0 \to \gamma k_0 v_\parallel$, and the canonical momenta are conserved so long as $P_x^2 + P_y^2 \ll p_\parallel^2$ (small δv_z).

An extensive comparison between orbital motion in the "realizable" and the "unrealizable" undulator has been undertaken by Diament [56] (who also included the constant axial magnetic field). He found that exact helical orbits can be obtained in the combined fields only for adequately small $B_\perp$ [i.e., small pump strength], contrary to the situation for the unrealizable undulator field alone [Eq. (5.10)]. The important parameter is $k_0 r_0$: this is the ratio of the transverse to axial electron velocities, while $(k_0 r_0)^2$ represents the energy division between transverse and axial components. A real undulator will give the electron motion comparable to the idealized undulator only if $k_0 r_0 = a_\omega / \gamma \ll 1$; this is ordinarily true for highly energetic beams, but it is one source of the gain-limiting requirement that $a_\omega \sim 1$. (In practice, $k_0 r_0 < 1$, because of limitations on the coil current.) Some insight into the condition $a_\omega \sim 1$ may be obtained from the fact that this condition also results in maximizing the single-particle undulator radiation on the axis.

Consideration of making a helical undulator from permanent magnet "building blocks" (see "The Linearly Polarized Undulator" in Section 5.4) has turned up a design concept known as the mutipole ring. A disk of rare-earth–cobalt magnetic material can be assembled of several individual dipole sections (or sectors), where the dipole moment rotates by a fixed amount from one sector to the next. The undulator is then made of an axial stack of these disks, each being rotated about the axis of the undulator relative to its neighbors [94].

5.4   Other Types of Undulators

In this section we shall tabulate and discuss other undulators, which, with one exception, are only infrequently used.

The Cusp Undulator

This is made up of circular rings of current or magnetic material. Mathematically, the magnetostatic solutions are

$$B_r = -B_\perp I_1(k_0 r)\cos k_0 z$$

$$B_z = +B_\perp I_0(k_0 r)\sin k_0 z \tag{5.22}$$

but one must bear in mind that these solutions arise from sinusoidal surface-current distributions. When the electron is far from the axis, the fields from the bifilar helical and cusp undulators are not very different. When $k_0 r_b > 1$, the electrons are closer to the windings and there is an exponential increase in field. This would permit some reduction in the period l_0, together with a shortening of the radiated wavelength for given γ; however, the electron motion has higher harmonic content, and—worse yet —the amplitude of the rippled field can vary substantially across the beam, as the field is less homogeneous off axis. All this is more relevant to the beam location than to the field configuration, but it is well to note that $k_0 r_b \ll 1$ is to be preferred in all cases. Since the bore diameter of the undulator is a few centimeters, this is also the range of l_0. Note that, in contrast with the bifilar helix, as $r \to 0$ we have $B_r \to 0$ but B_z is finite; and for $k_0 r_b > 1$, $B_r \approx B_z$. A filamentary beam on the axis of a cusp undulator will not respond to the FEL excitation.

The cusp undulator can be set up with ferromagnetic rings, and a version that operates with eddy currents has been studied by Jacobs [101]. In that instance a transient solenoidal magnetic field soaks through a set of conducting rings to produce both the undulator and the guiding field. Owing to the eddy currents or the magnetic material, the axial field is always less than its vacuum value.

Electrostatic Undulators

The use of an electrostatic field to pump the electron quiver motion is quite similar to the magnetostatic undulator in principle. A serious drawback is the size of the field required. To produce the same quiver velocity as a magnetostatic undulator which generates 1 kG, the electrostatic undulator would have to provide 300 kV/cm with the same periodicity. To date there has been little interest in developing such a device. One variation that might appear attractive is to use the electrostatic charge of the beam; this will create its own undulator if the radius of the drift-tube wall containing the beam is periodically rippled. Unfortunately, this also establishes a slow-wave structure which couples effectively to the beam and is a source of very powerful microwave radiation [142]. This radiation can backscatter from the electron beam, giving a type of "two-stage" FEL (see Chapter 7).

An unusual device which is suggestive of an FEL with a "natural" electrostatic undulator is the channeltron, where a highly energetic electron or positron beam is passed through a nearly perfect crystal between (for example) two atomic planes separated by the characteristic lattice spacing d. In the axial (z) direction, $d' = d/\gamma$, and therefore in that direction the crystal looks almost like a continuously charged string (of ions) to the passing electron. In the transverse (x) direction, the electron motion is periodic—for small amplitude—and the oscillation in the electrostatic potential well of the ions depends on the crystal potential $U = \frac{1}{2}kx^2$, the plane spacing, and the electron energy: $\omega_0 = (\gamma k/m)^{1/2}$. After Doppler-shifting the frequency of the radiation set up by this transverse motion to the laboratory, we get $\omega = (\omega_0/\gamma)(1 - \beta\cos\theta)^{-1} \approx 2\gamma^{3/2}(k/m)^{1/2}$, which is not quite the FEL scaling relation. Experimental work has been underway since 1979 [17] using thin crystals and very high-energy beams. There is a wide variation of possible frequencies, depending on the representation of the potential and the way the charge interacts with it, but in general it is found that the radiation has fair coherence, is many times the bremsstrahlung level, and has a photon energy in the range of several kilovolts.

Another "natural" undulator can be made from a stack of foils, such as Be and Li, arranged alternately. Transition radiation will be produced by the passage of an energetic electron through media with differing dielectric properties. A very high-energy beam is necessary to reduce the electron-ion scattering.

The Linearly Polarized Undulator

While this undulator can be constructed of current-carrying elements [157], recently very high quality blocks of permanent-magnetic material have become available, of which currently samarium-cobalt or NdFe is preferred. Long variable-parameter undulator systems can be fabricated for FELs where the guiding field is small or zero; the assembly is very amenable to rearrangement and reuse. The geometry is linearly polarized (Fig. 2.1). The field components satisfying $\nabla \times \boldsymbol{B}_\omega = 0$ for a sinusoidal winding are

$$B_x = 0$$
$$B_y = B_\perp \cosh k_0 y \cos k_0 z$$
$$B_z = -B_\perp \sinh k_0 y \sin k_0 z \qquad (5.23)$$

When a guiding field in the axial (z) direction is included, the motion is much more complicated than for the simple undulator treated in Section 2.2. The orbit can be obtained from

$$v_y = \frac{\Omega_0 \Omega_\perp v_\parallel}{\Omega_0^2 - \gamma^2 k_0^2 v_\parallel^2} \cosh k_0 y \cos k_0 z$$

$$v_x = \frac{\Omega_\perp \gamma k_0 v_\parallel^2}{\Omega_0^2 - \gamma^2 k_0^2 v_\parallel^2} \cosh k_0 y \sin k_0 z$$

$$z \approx v_\parallel t \qquad (5.24)$$

for approximately constant $v_\parallel$. The ratio of the maximum excursion in the x-direction to the excursion in the y-direction is

$$\frac{\Delta x}{\Delta y} = \frac{\gamma k_0 v_\parallel}{\Omega_0} \qquad (5.25)$$

and is a measure of the magnetoresonance condition. The polarization of the radiation is generally elliptical [163]. As a result of the radial asymmetry, there is a net drift of off-axis electrons [152], which is illustrated in Fig. 5.5. The orbit simulation includes the beam self-fields, and the orbit is shown projected in the x-y plane. The drift is a $\boldsymbol{B}_0 \times \nabla B_\perp$-type motion which does not occur for vanishing guide field.

When assembling a linearly polarized undulator from permanently magnetized elements, the following are important:

1. The configuration itself must not produce fields which will depolarize the magnetic material.
2. The beam must travel far from the edges of the magnets to avoid harmonic content in the orbit.
3. Provision must be made for varying the amplitude $B_\perp$ (varying the gap width).
4. The magnets must be selected and arranged so that random variations in dipole moment tend to cancel.
5. The magnets must be mounted on a firm carriage in such a way that they will not crack or move under the combined forces.

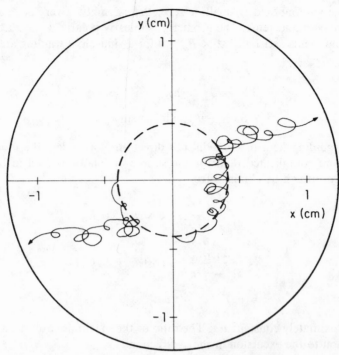

Figure 5.5 Orbit of an electron in a linearly polarized undulator and a guiding field, showing drift (transverse plane) [152]. © 1982 AIP.

The undulator used by the LANL FEL project [197] is representative of what can be done with a samarium-cobalt system. It used 314 permanent magnets, each $0.5 \times 0.5 \times 3.5$ cm in size. These were arranged into 40 periods covering 100 cm in length. The gap between the rows of magnets is 0.88 cm, and the peak magnetic field at the beam is 3.1 kG. There was a matching section, 5 cm long, at the entrance and exit, consisting of a pair of rotatable magnets which could be adjusted to provide a smooth transition from zero to full magnetic field.

The linearly polarized undulator, fabricated from permanent-magnetic blocks, is a popular choice for high-energy accelerator systems. As a rule, the helical undulators are superconducting, except where short-period operation has permitted the current to be pulsed through the magnet windings. The helical undulator provides better focusing properties, particularly for intense, low-γ electron beams, than does the linear dipole undulator [34].

5.5 Sources of Inhomogeneous Broadening

In quantum electronics, the spectral-line radiation is characterized by a homogeneous width and an inhomogeneous width. In a two-wave FEL, the former is the width of the spontaneous spectrum, broadened by transit-time effects (Chapter 2), while in the Raman FEL the homogeneous width is $\sim g$ (Chapter 4). Inhomogeneous broadening mechanisms operate on each electron individually, and upon combination result in an enhanced line width of the system; this reduces the ability of the beam to respond as a unit, and lower gain is one consequence. If the FEL gain spectrum is broadened homogeneously, oscillation will occur typically only on one resonator mode, located near the maximum gain; this follows because oscillation at a particular wavelength will reduce the gain for the entire line (see "Arbitrary Amplitude" in Section 3.3). For the inhomogeneously broadened line, oscillation simultaneously in several resonator modes may be possible.

There are several inhomogeneous-broadening mechanisms, and in this section we shall discuss only a few which seem to be common to most FELs. They have the property of depending upon where the individual electron happens to be or where it originated, and so result from orbital effects. Since the FEL radiation is observed along the axis in the direction of electron motion, the relevant variable is $(\delta\gamma/\gamma)_{\shortparallel}$, which is approximately the normalized distribution of parallel electron momenta or electron energies.

Electron-Beam Space Charge

The electron beam itself is one source of inhomogeneous broadening. Although the accelerator provides a fixed energy for each electron, $(\gamma - 1)mc^2$, the electrons move in a beam where the equilibrium is characterized by a distribution of electrostatic potential and transverse rotational velocity, both of which depend upon the radial location of the electron (Section 5.2). Thus the kinetic energy and, more importantly, the parallel electron velocity must vary throughout the beam. The spread in axial velocities is related to $(\delta\gamma)_{\shortparallel}$, as we have found already in Chapter 4, by

$$\left(\frac{\delta\gamma}{\gamma}\right)_{\shortparallel} = \frac{\gamma^2 \delta v_{\shortparallel}}{c} \tag{5.26}$$

We can estimate the momentum spread of the electrons in the beam using conservation of energy, Gauss's law to calculate the electrostatic potential, and the equilibrium treatment of Section 5.2 to determine the rotational velocity. The rotation increases as one moves radially outward in the beam, whereas the electrostatic potential decreases; these two effects partly compensate for each other, and for the rigid-rotor model with small rotational velocity, we find for the space-charge effects

$$\left(\frac{\delta\gamma}{\gamma}\right)_{\shortparallel,\,sc} = \frac{\omega_{p0}^2 r_b^2}{4\gamma c^2}\left(1 - \frac{2\omega_{p0}^2}{\Omega_0^2 \gamma}\right) \sim \frac{\nu}{\gamma} \tag{5.27}$$

For the Brillouin-flow limit, $\omega_{p0}^2 = \gamma\Omega_0^2/2$, the beam rotates faster at $\omega_\theta = \Omega_0/2\gamma$, and the compensation is exact. Yet it is difficult in general to arrange zero parallel velocity shear, except under conditions where the electrons are emitted from a cathode in zero magnetic field [159]. If the electron pulse is very short, additional velocity spread may result from the axial component of the space charge field.

The Undulator

Electrons in the beam have a certain distribution of momenta and are located at various radii. Once the beam enters the undulator, two related effects will cause additional inhomogeneous broadening in this environment.

First, $B_\perp$ will vary radially, so that the radius of the electron helical orbit is not constant, and the quiver velocity varies across the beam. An axial velocity shear arises [see Eqs. (5.13) and (2.10)]:

$$\delta v_\parallel \approx c(\beta_\perp k_0 r_b/2)^2 \tag{5.28}$$

and the equivalent axial normalized momentum spread across the beam profile is

$$\left(\frac{\delta\gamma}{\gamma}\right)_{\parallel,\,und} = \frac{b_\omega^2 r_b^2/2}{1 + b_\omega^2/k_0^2} = \frac{k_0^2 r_b^2 a_\omega^2/2}{1 + a_\omega^2}\left(< \frac{1}{2N}\right) \tag{5.29}$$

We have already noted that the orbit-size requirement $k_0 r_0 = a_\omega/\gamma \ll 1$ restricts the gain and places the limit $a_\omega \lesssim 1$. Now there is a new requirement on the size of the electron beam: we ask that the undulator inhomogeneous broadening be no more than the FEL homogeneous broadening: $(\delta\gamma/\gamma)_{\parallel,\,hom} \lesssim 1/2N$. From this follows $k_0 r_b < 1/N^{1/2}$. Since $N \sim 100$ and $k_0 \sim 1$ cm^{-1}, a typical electron-beam radius is about 1 mm. The contributions to $(\delta\gamma/\gamma)_\parallel$ from Eq. (5.29) and Eq. (5.27) are of opposite sign, that is, they tend to cancel.

Secondly, the electrons may have an intrinsic azimuthal motion not related to the undulator. There may be a good reason to consider situations where $\beta_\perp$ is large at the entry to the undulator (see "With Undulator" in Section 6.3), and this may become a very important source of inhomogeneous broadening [see Eq. (5.21)]. On the other hand, if one desires minimum inhomogeneous broadening, one should inject the electron beam into the undulator with momentum tangent to a spiraling line of force. A practical rule is to "taper" the pump field $B_\perp$ adiabatically over several periods l_0, and if there is a guiding field, this should extend over several cyclotron periods $2\pi\gamma v_\parallel/\Omega_0$ as well. Injection at magnetoresonance will result in beam loss or "heating" after a few periods.

Transverse Motion

In the accelerator, all electrons may emerge with the same energy, but—owing to the geometrical details of the diode or accelerating elements—there will be a distribution of momenta between parallel and transverse components:

$$\frac{\delta v_z}{v} \approx \frac{\delta v_x^2}{2v^2} \tag{5.30}$$

It is the practice in accelerator physics to characterize this by a parameter known as emittance (ε) or by the normalized emittance $\varepsilon_N = \gamma\beta\varepsilon$. In an ideal accelerator, ε_N is constant throughout the acceleration of the charge, but this is not the case in practice, because the transverse forces responsible for the off-axial motion are not altogether linear. For a beam with emittance ε_x, the maximum value of v_x/v is ε_x/x; for radially symmetric beams, Eq. (5.30) becomes [143]

$$\frac{\delta v_{\parallel}}{c} = \frac{\beta\varepsilon^2}{2r_b^2}$$

or

$$2\left(\frac{\delta\gamma}{\gamma}\right)_{\parallel,\varepsilon} = \frac{(\gamma\beta\varepsilon/r_b)^2}{1+a_\omega^2} = \frac{(\varepsilon_N/r_b)^2}{1+a_\omega^2} \tag{5.31}$$

Another parameter related to emittance is the beam brightness, the electron flux per unit solid angle, defined by

$$B_\varepsilon = \frac{I_b}{\pi^2\varepsilon_N^2} = \frac{J_B}{2\pi(\delta\gamma/\gamma)_{\parallel,\varepsilon_N}} \tag{5.32}$$

where I_b and J_B are the current and current density. As I_B is increased, it is important to keep in mind that we also require $v/\gamma \ll 1$. We have already observed that the gain of the FEL will not be reduced by effects of inhomogeneous beam momentum spread as long as $(\delta\gamma/\gamma)_{\parallel} < 1\% \sim 1/N$ (a quantitative demonstration of this, applicable to both two- and three-wave FEL phenomena, is provided in Chapter 6). Therefore the brightness of an electron beam should scale roughly as the current density, provided the beam meets our simple energy-spread criterion. Intuition suggests that a brighter electron beam must be good for something (e.g., greater radiation), since the beam power scales as J_b, but exactly what is it that is being optimized?

The brightness of the laser radiation (defined as the power per steradian per hertz of bandwidth) from a simple FEL can be related to the electron-beam brightness formula above, using the following argument. The power emitted by the FEL amplifier is $\eta I_b V_b$, where $\eta \sim 1/2N$, $V_b \sim \gamma mc^2$; the bandwidth is $\sim 1/N$. The radiation spreads out from the FEL aperture with angle $\Delta\theta_0 = 1/\gamma N^{1/2}$ (given in Chapter 3), so the solid angle of radiation is $\pi(\Delta\theta_0)^2$, and the laser brightness B_L is roughly $I_b \gamma^3 mc^2 N/2\pi$. Now we turn to the matter of the electron-beam brightness. The normalized beam emittance should be chosen so that it presents negligible inhomogeneous broadening in the two-wave FEL where the homogeneous broadening is already $(\delta\gamma/\gamma)_{\parallel, \text{hom}} \sim 1/2N$. This implies that the electron divergence $v_\perp/v < (2\lambda_s/l_0 N)^{1/2}$, or $\varepsilon_N < r_b/N^{1/2}$. Then, using Eq. (5.32), the electron-beam brightness is $B_\varepsilon \approx I_b N/\pi r_b^2$. Taking the ratio, we have

$$\frac{B_L}{B_\varepsilon} \sim \gamma^3 mc^2 r_b^2 \sim \frac{\gamma^3 mc^2}{k_0^2 N} \tag{5.33}$$

A bright FEL optical beam is simply related to a bright electron beam. The laser brightness is a distinct quality which can be optimized by design, rather than the efficiency. The importance of electron-beam brightness suggests which accelerators may find application as FELs. The FEL gain scales linearly with beam brightness.

The above calculation illustrates the interplay of the various criteria used to establish the FEL parameters. For example, the diffraction angle $\theta_d \sim \lambda_s/r_b$ turns out to be $\Delta\theta_0$ (see "Large Amplitude—Approximate Result" in Section 3.3) once the undulator length is set equal to the Rayleigh range. A more conservative choice might be to reduce θ_d by increasing r_b. However, recalling Eq. (5.29), we have $a_w k_0 r_b < 1/N^{1/2}$. But then we find a restriction on a_ω: $a_\omega < \gamma/\pi N$.

Commentary

An empirical formula which relates emittance to beam current is the Lawson-Penner formula

$$\varepsilon_N = sI_b^{1/2}(\text{kA}) \text{ cm-rad} \tag{5.34}$$

where S is a factor of order 0.1–0.3 for linear-accelerator configurations using a thermionic cathode (≤ 10 A/cm^2) in zero magnetic field. Some of the emittance in the rf linac has to do with the bunchers, which compress the peak current in the micropulses. A typical emittance— ~ 100 mm-mrad —can be reduced perhaps an order of magnitude by careful attention to this point (an rf feedback control system to stabilize the rf power to the buncher cavities is helpful [197]). In the example of a low-current, DC beam, it is possible [67] to reduce the emittance to a point where it is dominated by the thermionic emission from the hot cathode, i.e., $\sim 2r_b(kT/mc^2)^{1/2}$.

High-energy, low-current beams have relatively little trouble in reaching much lower emittance—$\approx \pi$ mm-mrad is a typical number—and they are very well suited for more sophisticated FEL applications.

As the beam is transported through the FEL system, it is useful to follow the beam profile using an envelope equation, which can be obtained by doctoring Eq. (5.5). We study the radial motion of an electron at the beam boundary, r_b; this is a nonequilibrium case, so the left-hand side of Eq. (5.5) contains the radial inertia term $\gamma m \ddot{r}_b$, rather than zero. On the right-hand side, set $B_0 = 0$ and introduce the emittance in place of the v_θ term. Then

$$\frac{d^2 r_b}{dz^2} = \frac{2eI_b}{\gamma^3 m v r_b} + \frac{\varepsilon_N^2}{(\gamma\beta)^2 r_b^3} \tag{5.35}$$

where the first term on the right represents the net effect of the space-charge forces. Where external focusing elements are involved, their effect can be modeled by adding a term $f^2(z)r_b$ to the left-hand side of Eq. (5.35). If there is no focusing and we neglect space charge effects, then Eq. (5.35) is easily integrated and one finds that the electron beam—once focused to a diameter r_b—will double its area in traveling an axial distance $\sim r_b^2 \gamma / \varepsilon_N$. Choosing typical rf linac parameters (say $r_b \sim 1$ mm, $I_b \sim 1$ A), the Lawson-Penner relation shows that the beam will only travel about 200 cm before spreading out considerably. Does this define the half-length of the undulator? Not necessarily, as the undulator may have some focusing property or a guiding magnetic field may be used to channel the beam. It is interesting that the spreading of the electron stream matches the diffractive spreading of the optical beam if $\varepsilon_N \sim \gamma \lambda_s$ (usually a more exacting requirement than Lawson and Penner's].

When a constant magnetic field is used to transport the beam, the focusing term on the left side of Eq. (5.35) is just $r_b/(2r_{ce})^2$, where r_{ce} is the electron gyroradius at B_0 [143]. Setting $\ddot{r}_b = 0$, one obtains the equilibrium radius of the electron beam in either the emittance-dominated or the space-charge-dominated regime: the dependences on B_0 are not the same. A guiding field is essential when the space-charge effects are important. A very strong magnetic field will also reduce the transverse motion due to fringing components of the electric field in the diode; this is done by reducing the gyroradius of the electron so that it samples very little potential change in the direction transverse to the axis. The normalized parallel energy spread scales as $\gamma^2 \beta_\perp^2 / 2 \propto V_d^2 / B_0^2$.

Research on high-current-density, cold-cathode, magnetized diodes has shown that S can be reduced to roughly 0.04 if one is willing to dispense with a large part of the initial beam by aperturing [151]. A value of $\beta_\perp < 0.050$ was reported [100] from an apertured-beam experiment at NRL; the parallel velocity spread was $< 0.1\%$. Recently, Sheffield et al. [171] have used a magnetic field ~ 100 kG to propagate a cold [$\beta_\perp < 0.030$, $(\delta\gamma/\gamma)_{\parallel}$

< 2.5%], intense [0.4 MA/cm^2, $\nu/\gamma < 1$, $\gamma = 7.4$] electron beam. In the latter case, although the emittance is satisfactory for FEL applications, the space charge is too high. At this point it is still difficult to transport an intense beam into the undulator in any geometry using a guide field which is more complicated than uniform field lines.

Effect of Inhomogeneous Broadening on Gain

To avoid loss of gain, one endeavors to arrange that the inhomogeneous broadening is less than the homogeneous broadening. A typical set of numbers for a two-wave FEL might be: $N = 100$; $\varepsilon_N = 30$ mm-mrad, $r_b = 1$ mm, or $(\delta\gamma/\gamma)_{\|,\varepsilon} \approx \frac{1}{2}\%$; $b_w \sim 1$, or $(\delta\gamma/\gamma)_{\|,und} \approx \frac{1}{2}\%$; $(\delta\gamma/\gamma)_{\|,s.c.} \sim \nu/\gamma \ll 1$. The various sources of inhomogeneous broadening can be combined as small independent terms, as the RMS sum of the individual broadenings. Since the line width $\delta\omega/\omega$ is proportional to the energy broadening $(\delta\gamma/\gamma)_{\|}$, the total line width is

$$\left(\frac{\delta\omega}{\omega}\right)^2_{total} = \left(\frac{\delta\omega}{\omega}\right)^2_{hom} + \sum_i \left(\frac{\delta\omega}{\omega}\right)^2_{i,inhom} \qquad (5.36)$$

where the various inhomogeneous terms have been already discussed in the first three subsections. The homogeneous term is, as we have shown in Chapter 3, roughly $1/N$ wide; the gain formula, Eq. (3.30), already incorporates homogeneous broadening, since it can be factored as $nB^2_\perp$ $(\lambda_s l_0)^{3/2} LN^2 \propto (\delta\omega/\omega)^{-2}_{hom}$. However, Eq. (5.36) assumes that the spontaneous radiation intensity will be inhomogeneously broadened in a Gaussian way. The quantitative calculation of the resulting FEL gain, obtained in the customary way be taking the derivative of the inhomogeneously broadened spontaneous spectrum, has been reported by Renieri [160] and Ciocci et al. [44]. The gain falls in a monotone, but complicated, way with increasing inhomogeneous broadening, and if the latter is large, the gain decreases [47] approximately as $[4\pi N(\delta\gamma/\gamma)_{inhom}]^{-2}$. Since high-energy, low-current accelerators provide very high-quality beams, ordinarily homogeneous broadening dominates the inhomogeneous terms. The reader is reminded that if a nonuniform undulator is used, the homogeneous broadening may be larger than $1/N$ [Chapter 3, below Eq. (3.36)].

In the Raman FEL, a similar qualitative conclusion obtains, but it is necessary to model the inhomogeneous terms in the theory using a broadened distribution of electron momenta (Section 6.2). The various inhomogeneous terms are gathered together into a representation of the electron momentum distribution, used in a kinetic theory of the FEL. The situation with the Raman FEL is quite the opposite to the FELs which use high-energy electron beams: inhomogeneous broadening is an important effect, and may cause a significant deterioration of gain (see "Zero Guiding Field" in Section 6.2). Furthermore, there may be a degree of correlation between the various inhomogeneous terms owing to large-amplitude motions. This is particularly true for the situation near magnetoresonance, where numerical simulation of electron orbits [89] has shown a very complex interaction between the initial transverse motion of the electron, the undulator field, and the length of time over which the interaction is maintained.

The space-charge contribution to the inhomogeneous broadening becomes appreciable in the Raman FEL. Assuming that the inhomogeneous terms given in the first three subsections are small and combine quadratically—an *Ansatz* open to question in view of what was said above—one can calculate the value of r_b which minimizes the inhomogeneous broadening. The result (see [152]) is

$$r_b^2 \sim \frac{\varepsilon_N}{\left| b_\omega^4 + \left(\omega_p/2c \right)^4 \right|^{1/4}} \sim \frac{\gamma \varepsilon}{k_0 a_\omega} \qquad (5.37)$$

which gives an optimum beam radius ~ 1 mm under typical conditions. This radius is also practical for reasons not having to do with beam quality: to wit, beam equilibrium and diffraction. Taking the usual approximation $a_\omega \sim 1$, as well as the limit of space-charge effects unimportant, Eq. (5.37) will relate the beam area to its emittance: $A_b \approx \varepsilon_N l_0/2$.

Diagnostics

Measurement of the momentum spread in high-energy, low-current beams is done with a standard laboratory diagnostic, the electron spectrometer. For intense beams, less is known, because of the complicating factor of lower energy and high space charge. An analysis done several years ago by Friedman et al. [79] showed that magnetic measurements could be used to determine $v_\perp$ and estimate a limit for the energy spread; others have estimated $v_\perp$ from determinations of beam size or scrapeoff in a given magnetic field. Talmadge et al. [189] used a calibrated thin foil X-ray bremsstrahlung method to determine the $v_\perp/v_\parallel$ of a beam emitted by a foilless diode. It was found that $v_\perp \sim B_0^{-1}$. Shefer et al. [173] have reported measurements of $v_\parallel$ and $v_\perp$ by observation of the radial electrostatic beam potential as well as the cyclotron wavelength.

A potentially useful noninteractive diagnostic has been developed at Columbia by Chen and Marshall [43], who analyzed the spectral distribution of wavelengths in the visible ($\sim \frac{1}{2} \mu$m) which arise from Thomson backscattering of a pulse of TEA CO_2 laser radiation at 9.6 μm from an intense beam ($\gamma = 2.3$, $j = 10^3$ A/cm^2). Equations (2.18) and (2.20) may be used to interpret the experiment. The scattering cross section is enhanced by $4\gamma^2$ in the backscattered direction with respect to the electron motion.

The spectrum of radiation observed in the laboratory frame can be related to a narrow Gaussian in the rest frame, using theory by Zhuravlev and Petrov [200], reinterpreted later by Kukushkin [111]. Analysis of the spectral width of the scattered light revealed $(\delta\gamma/\gamma)_{\|,\text{inhom}} \sim \frac{1}{2}\%$, showing that intense, comparatively low-energy electron beams are suitable for Raman FEL applications. In this configuration, the beam space charge was the dominant broadening term. When the undulator was used, the total inhomogeneous broadening increased, and the data showed that the space charge and undulator [Eq. (5.29)] contributions should be combined quadratically (Fig. 5.6).

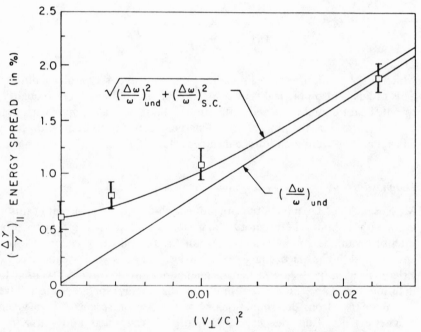

Figure 5.6 Measured $(\delta\gamma/\gamma)_{\|}$, using Thomson Backscattering diagnostics, showing the dependence on $v_\perp/c$ [eq. 5.18] induced by the undulator. The data show a quadratic combination of the inhomogeneous broadening from the space charge and undulator effects. Also evident is the quadratic dependence of undulator broadening on $(v_\perp/c)$. (Chen, 1984).

6

Electron Motion and FEL Theory; Slow-Wave FELs

6.1 *Introduction*

This chapter contains a miscellany of FEL-theory topics which either expand upon the development in Chapters 3 and 4 or merit discussion in their own right. (The latter include the stimulated Čerenkov FEL, a slow-wave device which uses no undulator; we discuss this in Section 6.4.) With the additional background provided by the orbital description of Chapter 5, there is motivation to incorporate systematically a detailed description of the electron motion in the theory, striving for better accuracy.

In Chapter 4 we undertook a theoretical treatment of the FEL using a traveling-wave amplifier approach; both two- and three-wave effects were incorporated in this model. The next step in the theoretical development is to formulate a boundary-value problem, replacing the cold-fluid treatment with a kinetic one which will include the effect of a distribution of electron momenta. It is no surprise then that the calculation becomes quite complex, requiring numerical analysis; nevertheless the results are most informative with regard to the interplay of two- and three-wave effects, especially when the electron thermal spread is substantial. Unfortunately, experiment usually provides very limited information about the details of the kinetic distribution. We shall outline the theory and quote a few important examples [Section 6.2].

The presence of an axial guiding field adds further complications, as one can now appreciate, having seen the complex influence of this magnetic field upon the orbital motion (Chapter 5). When the unperturbed component of motion transverse to the guiding field becomes important, the FEL physics takes on a different character—somewhat resembling the electron cyclotron maser, but differing in that the radiation is considerably Doppler-upshifted by the large parallel electron-velocity component. We shall survey undulator-free and cyclotron-undulator hybrid FELs in Section 6.3.

6.2 *The FEL as a Traveling-Wave Amplifier*

Many new features were incorporated into the traveling-wave amplifier theory of the FEL by Bernstein and Hirshfield [26]. They included equilibrium potentials which were exact to all orders of the pump, and used a set of basis vectors which simplified the problem. They used the canonical undulator [Eq. (5.15)] and neglected radial dependences. An important feature was the formulation of a boundary-value problem recalling the traveling-wave amplifier, which—although involving different physics—has the common characteristic of propagating several modes at once. This permits a calculation for the coupling loss, i.e., the division of power at the FEL input between the different modes, owing to the boundary condition on field amplitudes there. Their treatment included kinetic effects, and this aspect was pursued in numerical detail by Johnston [102] and Ibanez and Johnston [99]. Later analysis [77, 27] included a guide magnetic field.

We shall very briefly review this theory, without going into the details of the analysis, which for full understanding requires a detailed treatment of inverting the Laplace transform using methods of analytic continuation by deformation of contour. A kinetic treatment requires solution of the relativistic, linearized Vlasov equation (collisionless Boltzmann equation)

$$\frac{\partial f^{(1)}}{\partial t} + \dot{z}_{(0)} \frac{\partial f^{(1)}}{\partial z} = \frac{|e| \boldsymbol{E}^{(1)} \cdot \boldsymbol{u}}{mcu} \frac{\partial f^{(0)}}{\partial u} \tag{6.1}$$

where the superscript (1) indicates the first-order perturbed quantities and the superscript (0) indicates unperturbed quantities (the perturbation is the field $E^{(1)}$ of the scattered wave). The orbits of the electrons in the undulator enter the problem, as they are the solution of the unperturbed situation. The zeroth-order axial velocity is $\dot{z}_{(0)}$, and is normalized as $u_{(0)} = \gamma \dot{z}_{(0)}/c$. Maxwell's equations for potentials determine the EM fields:

$$\frac{\partial^2 A^{(1)}}{dz^2} - \frac{1}{c^2} \frac{\partial^2 A^{(1)}}{\partial t^2} = -\frac{4\pi}{c} j_\perp^{(1)} \tag{6.2}$$

$$\frac{\partial^2 \Phi}{\partial z \, \partial t} = 4\pi j_z^{(1)} \tag{6.3}$$

where the current density is

$$j^{(1)} = -|e| \int d^3u \, \boldsymbol{u} \frac{cf^{(1)}}{\gamma} \tag{6.4}$$

The helical electron motion in the ideal undulator enters the theory via the zeroth-order, or equilibrium, electron distribution function $f^{(0)}$. This can be factored as

$$f^{(0)} = n\delta(\beta_1)\delta(\beta_2)g(\boldsymbol{u}) \tag{6.5}$$

where

$$\beta_1 \equiv u_x - \frac{|e|A_{x\omega}}{mc^2}$$

$$\beta_2 \equiv u_y - \frac{|e|A_{y\omega}}{mc^2} \tag{6.6}$$

are the normalized canonical momenta, the constants of the motion in the idealized helical undulator. The analysis proceeds in general by Laplace transformation, solution for the dielectric tensor, calculation of the dispersion relation, and inversion of the Laplace transform to calculate the evolution of the field amplitudes along the undulator.

Zero Guiding Field

In their first paper, Bernstein and Hirshfield [26] chose a cold (delta-function) form for $g(u)$ and calculated several cases appropriate to the early Stanford FEL amplifier results. Figure 6.1 shows one of their calculations for gain versus wavelength, for a low-energy (2 MV), intense (6 kA/cm^2) beam, operating as an FEL with $k_0 = \pi$ cm^{-1} and a rather strong pump ($a_w = 0.47$). Two cases—for an undulator of 50-cm length ($N = 25$) and 100-cm length ($N = 50$)—are shown, illustrating two main points. First, although one is within the regime of complex roots (exponentially growing convective instability), there is clear evidence for the spatial interference of the modes. As the interaction continues for a greater length of undulator, the exponentiating mode dominates the others. Secondly, the power gain is not what would be expected if the signal present at the FEL input merely grew exponentially according to Eq. (4.54). What actually happens is that the input signal amplitude is divided among the several modes which can be excited, and the amount available for the exponentially growing mode is reduced by a factor of about 6, -15.5 dB in power. [This is a familiar effect in traveling-wave-tube analysis, where the input power is divided between three propagating modes, so that each one has one-ninth of the input power.

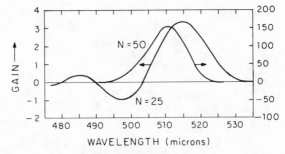

Figure 6.1 Gain of signal in an undulator of 25 or 50 periods, showing combined exponential and interference effects [26]. © 1979 APS.

The exponentially growing signal then grows not as e^{az} but as $\approx e^{az}/9$ (Fig. 6.2). The factor of $\frac{1}{9}$, or -9.6 dB, is referred to as the *coupling loss*.]

Ibanez and Johnston [99] represented $g(u)$ as a Gaussian distribution, where the "thermal parameter" is the parallel momentum spread. Taking a weak-pump example ($a_\omega = 0.06$) for a dense-beam ($\gamma = 2.5$) Raman FEL amplifier at 1.2 mm, they found only a modest drop in the spatial growth coefficient until the momentum spread increased (Fig. 6.3) to about 2%, followed by a more rapid decline. The coupling loss, about -12 dB, decreased as the momentum spread increased, which showed that some of the modes merged and lost their identity as the beam "temperature"

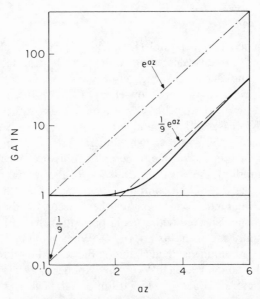

Figure 6.2 Growth of power along a traveling-wave amplifier, showing coupling loss.

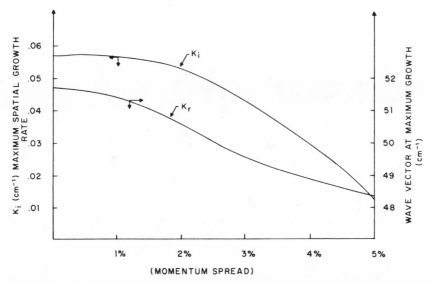

Figure 6.3 Dependence of wavenumber and spatial growth rate upon normalized parallel momentum spread [99]. © 1983 IEEE.

increased. A study of power growth along the undulator showed (Fig. 6.4) that both the spatial-interference and exponentially growing modes damp as the momentum spread is increased, but it is not until one reaches a characteristic length

$$L_\delta = \frac{l_0}{(\delta\gamma/\gamma)_{\shortparallel}} \tag{6.7}$$

that the effect of thermal spread becomes important. Thus a momentum spread of 2% would imply that thermal effects on the beam would not become damaging as long as the undulator was no more than 50 periods long. Similar conclusions were obtained, using a simpler kinetic model, by Fruchtman and Hirshfield [81].

The results of the Ibanez calculation provide some insight into the design of a high-gain single-pass FEL amplifier. Suppose we accept $N = 50$ and a 2% total parallel momentum spread, taking $k_0 = 5.0$ cm^{-1}. From Fig. 6.4 the signal power appears to be amplified a factor ≈ 40. However, a real amplifier will have a finite filling factor f, and the gain coefficient of the exponential part of the signal scales as $f^{1/2}$, where in a typical situation $f \sim \frac{1}{2} - \frac{1}{10}$. The temptation to increase the undulator amplitude is restrained by the limitation on the momentum spread that can be induced by this element [Eq. (5.29)], given the ambient inhomogeneous broadening from the beam space charge [Eq. (5.27)]. These broadening terms vary as $B_\perp^2$ and ω_p^2 respectively, whereas the Raman gain coefficient scales as $\omega_p^{1/2}B_\perp$. It is therefore best to increase $B_\perp$ at the expense of ω_p.

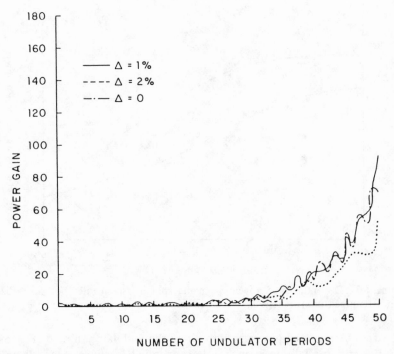

Figure 6.4 Power gain along an undulator, with momentum spread (Δ) as a parameter [99]. © 1983 IEEE.

The discussion above also permits us to reach some qualitative conclusions about the bandwidth of this FEL. We are considering a case where the power grows nearly exponentially along the undulator in the unsaturated case. As the length is increased, for fixed growth, the power spectrum will narrow (a similar situation occurs for an atomic laser: see [83]). Since increased electron-beam energy spread causes a drop in gain, it will cause an increase in power spectral width, which can be classified accordingly as a type of inhomogeneous broadening. Also, saturation of the FEL gain will cause an increase in the spectral bandwidth.

Finite Guiding Field

An effect of the guide magnetic field is to introduce many complications in the FEL analysis, most of them occurring in the vicinity of magnetoresonance, where electrons execute orbits with large transverse velocities and the theoretical basis is least accurate. Freund [74] has calculated the dependence of FEL gain on the axial magnetic field in the regime of low γ and high space charge, shown in Fig. 6.5. The idealized orbit model of Friedland was used, injecting the electrons into the undulator through a tapered region.

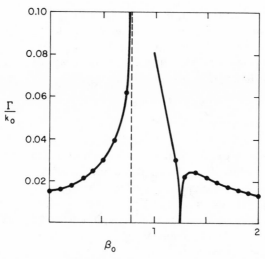

Figure 6.5 Enhancement of FEL gain by a guide magnetic field [74]. © 1983 APS.

The parameter β_0, defined as $eB_0/k_0mc^2\gamma$, describes the interaction, having chosen $a_\omega = 0.05$, $\gamma = 3.5$, and $\omega_p/k_0c = 0.1$. The guide field causes an enhancement of ponderomotive potential and an increase of FEL gain for the type I orbits ($\beta_0 < 0.7$ in this example), while the efficiency increases only modestly in the neighborhood of magnetoresonance. Although gain is also increased for the type II orbits, the FEL frequency is substantially diminished and its spectrum is broad. Farther from the magnetoresonance or the orbital instability points, some gain enhancement is possible, while enough $\beta_\parallel$ is retained so that the Doppler upshift of the FEL frequency is large. Experiments (Chapter 7) have shown that indeed the gain is enhanced, but close to magnetoresonance it drops precipitously, perhaps due to enhanced inhomogeneous broadening effects resulting from the large quiver motion.

McMullin and Davidson [137] have argued that there is competition between the FEL and cyclotron maser instabilities near magnetoresonance, and that the choice of an appropriate undulator length can bring out either effect or some combination. They find the FEL contribution dominates in the tenuous-beam, low-gain limit if

$$\left(k_0 v_\parallel - \frac{\Omega_0}{\gamma}\right)\frac{L}{2v_\parallel} = -\left(\frac{B_\perp}{2B_0}\right)^{2/3}\frac{L\Omega_0}{\gamma v_\parallel} \approx p\pi \qquad (6.8)$$

To summarize the matter of guide magnetic field, its use is a convenience at high energy and a necessity for intense lower-energy beams, at least in the context of beam propagation and equilibrium. It offers some promise for enhancing gain, but the difficulty is that the electron orbits can be very complex and this suggest that many competing effects will occur. Far from magnetoresonance, the theoretical treatment neglecting B_0 is probably accurate, providing the correct quiver velocity is used in the calculation. In the next section we shall see that the guide field offers some attractive possibilities for new FEL varieties.

6.3 FEL-Cyclotron Hybrids

From the kinetic theory of the previous section, it is clear that the radiation emitted by the FEL depends on the nature of the electron orbit. This follows directly from the method used to solve the kinetic equation, which involves a time integration along the unperturbed electron orbits (viz., the orbit that occurs in the absence of the radiation field). Given a specific orbit, we get a result. If we omit details of the orbit, the calculation will fall short of what is actually seen. Thus far, in FEL research, there has been good correlation between analysis and experiment. Until now the main constituents of the orbital motion have been the axial, relativistic streaming motion and the quiver motion induced by the undulator. This model is particularly good for the high-energy FELs. We now consider cases where there is an appreciable gyromotion of the electron around the guiding-field lines—either without or with an undulator. This complication to the motion means that not only is the analysis more involved, but we can be less certain of its reliability.

Without Undulator

The example with no undulator will be discussed first. The cold electron distribution function can be described by

$$\frac{n}{2\pi p_{\perp 0}}\delta(p_{\perp 0} - \gamma m v_{\perp 0})\delta(p_{\parallel} - \gamma m v_{\parallel}) \qquad (6.9)$$

A beam with large $p_{\perp 0}$ and negligible $p_{\parallel}$ is used by the gyrotron device, or electron cyclotron maser. Lacking a Doppler upshift, it is not a free-electron laser; also, the frequency is approximately the electron cyclotron frequency, Ω_0/γ. Given restrictions on the size of the guiding field as well as the practical difficulties of operating on higher harmonics, this device operates only in the microwave-millimeter spectral region, i.e., $\lambda > 1$ mm. Suppose instead we consider a device where both $v_{\parallel}$ and $v_{\perp 0}$ are large, in which event the electron will spiral around the lines of guiding field with pitch $2\pi\gamma v_{\parallel}/\Omega_0 \equiv \lambda_c$. This can be compared with the helical orbit (pitch l_0) in the undulator case. Since the FEL relationship is $\lambda_s \approx l_0/2\gamma_{\parallel}^2$, then

$$\lambda_s = \frac{\pi\gamma v_{\parallel}}{\gamma_{\parallel}^2 \Omega_0} \qquad (6.10)$$

This situation has been studied on several occasions during the last few years, with different objectives in mind. Ott and Manheimer [146] and Hirshfield et al. [97] showed that Doppler-shifted cyclotron radiation could be emitted from relatively dense beams, and they found a threshold $\beta_{\perp 0} \sim 1/\gamma$ for instability; far above threshold, the instability grows at a rate $\sim \beta_{\perp 0}\omega_p$. The source of free energy for this type of FEL instability is the transverse gyromotion of the electrons. The radiation and gain are broadband, since no periodic external device is present to fix the operating wavelength. Fruchtman and Friedland [82] have shown that a more asymetrical momentum distribution in space is favorable to higher growth. Recently, using computational models, Lin [123] has found potential for an efficient, high-power operating point, provided $\gamma \approx 5$. If one restricts the operating parameters of a compact device to $B_0 < 30$ kG and $3 \geq \gamma \approx \gamma_{\parallel}\sqrt{2}$ (since substantial transverse motion is needed for appreciable gain), then $\lambda_s \geq 1$ mm. Several megawatts at 2–4-mm wavelength has been obtained from devices based upon this principle using a pulsed, high-intensity electron beam [39].

Ride and Colson [162] have studied the case of a tenuous high-energy beam, using a test-particle approach similar to their FEL theory. They note that in order to stimulate emission from many randomly placed classical electron oscillators, the oscillators must be anharmonic [36]. In the relativistic limit the electron gyrofrequency does of course depend on the electron energy and is therefore anharmonic. The simplicity of this type of FEL suggests that it may occur in nature, in astrophysical situations.

With Undulator

When an undulator is added to this situation, we note that Eq. (5.20) predicts two Fourier components, Ω_0/γ and $k_0 v_{\parallel}$. An orbit simulation by Grossman [89] shows this is indeed the case (Fig. 6.6), where electrons are injected into the undulator with prior gyromotion. Possibilities for applying this to a new type of FEL were analyzed by McMullin and Bekefi [135], who took the situation of a filamentary beam on the axis of a cusp undulator, so that the pumping is driven by the longitudinal component of the undulator. Independently, Grossman and Marshall [92] analyzed and demonstrated a very similar case where the undulator could provide transverse pumping, but required that the EM fields have a longitudinal component (waveguide fields). The frequency of oscillation is given in either example by the intersection of the Doppler-shifted fast cyclotron mode

$$\omega = (k + k_0)v_{\parallel} + \frac{\Omega_0}{\gamma} \tag{6.11}$$

which is unstable for sufficiently large transverse motion, with the light line

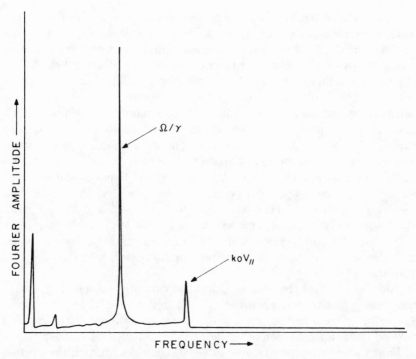

Figure 6.6 Fourier spectrum of electron orbital motion in an undulator, where the electron has appreciable initial transverse Larmor motion [92]. © 1983 IEEE.

$\omega_s = kc$, or

$$\omega_s = 2\gamma_{\parallel}^2 \left(k_0 v_{\parallel} + \frac{\Omega_0}{\gamma} \right) \qquad (6.12)$$

Actually, in view of the large electron orbital diameter, radiation at harmonics of the cyclotron or undulator frequency $2\gamma_{\parallel}^2 (pk_0 v_{\parallel} + q\Omega_0/\gamma)$ occur, and this may become technically important if the radiation can be directed into a high harmonic by design: a lower-energy beam could be used.

Theory (e.g., [90]) proceeds by a Vlasov integration of the unperturbed electron orbits [Eq. (5.20)] under the influence of the perturbing EM fields. This yields the current which sustains these fields, and an instability results under appropriate conditions; the threshold and growth rate depend on the product

$$\left(\frac{k_{\perp} v_{\perp}}{k_0 v_{\parallel}} \right) \left(\frac{k_{\perp} v_{\perp 0}}{\Omega_0/\gamma} \right) \qquad (6.13)$$

where $k_{\perp}$ is the waveguide cutoff. The convective-instability threshold is lowered by the presence of the undulator, and typical parameters sufficient for strong growth and high gain at $\lambda_s = 1.5$ mm are $\beta_{\perp 0} = 0.4$, $v_\perp/c = 0.1$, $\gamma = 2.5$. Thus, although substantial energy has been invested in the transverse component of motion, enough is reserved for the parallel component to give a substantial FEL-type upshift. Unlike the conventional FEL, the source of free energy for the instability is the initial gyromotion of the electrons ($v_{\perp 0}$), and in this respect the device is similar to the gyrotron. McMullin, Davidson, and Johnston [136] have shown that the use of a variable-parameter undulator will improve the efficiency of this type of FEL —which is otherwise rather inferior to the Raman FEL [134]. The conversion of a substantial fraction of the electron kinetic energy into spiraling gyromotion does make energy recovery less attractive as a means of overall system efficiency improvement.

6.4 Slow-Wave FELs

In the conventional (magnetobremsstrahlung) FEL, the interaction between the light wave and the slower space-charge fluctuation is made possible by the Doppler upshift of the space-charge wave by the undulator contribution $k_0 v_\parallel$. Another type of FEL uses no undulator, but instead slows the light wave so that it will couple to the beam wave. These are the Čerenkov and Smith-Purcell FELs. They resemble the traveling-wave tube, also a slow-wave device, where the EM mode and the plasma wave are coupled via the component E_z of the electric field.

Čerenkov FEL

The basic Čerenkov effect involves the radiation of a superluminous charge as it moves through a dielectric: $v > c/\mu$. The process generates a small amount of power, which is characteristic of the spontaneous radiation of a single charge. The radiation is emitted from the electron beam at the Čerenkov angle, $\theta_C = \cos^{-1}(1/\beta\mu)$. About twenty-five years ago attempts were made to increase the power by pre-bunching the electrons in the beam; however, there was no reinforcement of the bunching by an internal feedback mechanism, and the power remained small [45]. The development of higher-intensity, relativistic beams has changed this situation. Strong radiations from the stimulated Čerenkov process were observed first at Columbia [194], and an extensive research program at Dartmouth [193] has demonstrated that this mechanism will be an important source of radiation in the millimeter and submillimeter region, with excellent prospects for extension to short wavelengths as well [61].

In a Čerenkov FEL, a dielectric resonator is located near the electron beam; only if the beam energy is rather high and the dielectric is gaseous is it likely that the electron can pass through the medium directly, in view of the high beam intensity. The proximity of the dielectric lowers the speed of the radiation, permitting it to couple synchronously with the space-charge beam wave. As the beam propagates slightly faster than the radiation, it will bunch in the region of the retarding field, and thereby work is done on the field, which causes the bunching to increase. In Fig. 6.7 is shown the interaction of a light wave (drawn as a waveguide mode) and the beam space-charge wave. No intersection is possible until the light wave is slowed —it becomes asymptotic to the phase velocity c/μ for large k_z. Inspection of Fig. 6.7 shows that the frequency of the Čerenkov radiation becomes high as $v_{\parallel} \to c/\mu$.

A straightforward derivation has been provided by Walsh [196] for the case of the strongly magnetized beam (one-dimensional motion). Taking an rf field $\tilde{E}_z$, the electron will move according to

$$\frac{d\tilde{v}_z}{dt} = -\frac{|e|\tilde{E}_z}{m\gamma^3} \tag{6.14}$$

Hence

$$\tilde{v}_z = -\left(\frac{ie}{m\gamma^3}\right)\frac{\tilde{E}_z}{\omega - k_z v_{\parallel}} \tag{6.15}$$

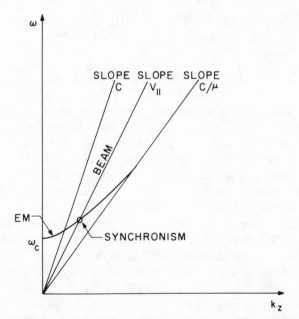

Figure 6.7 Slow-wave synchronism with the space-charge wave: the stimulated Čerenkov effect.

The linearized equation of continuity gives the density modulation

$$\tilde{n} = \frac{n_0 k \tilde{v}_z}{\omega - k_z v_{\parallel}} \tag{6.16}$$

Combining Eqs. (6.15) and (6.16), we get the current

$$
\begin{aligned}
\tilde{J}_z &= -n_0 e \tilde{v} - \tilde{n} e v_{\parallel} \\
&= \frac{i\left(\omega_p^2/4\pi\gamma^2\right)\omega\tilde{E}_z}{\left(\omega - k_z v_{\parallel}\right)^2}
\end{aligned} \tag{6.17}
$$

The wave equation is driven by $\partial \tilde{J}_z/\partial t$, and from this is obtained the dispersion relation

$$\left[\frac{\omega^2}{c^2} - k_z^2 - k_\perp^2 - \frac{\omega_p^2}{\mu^2\gamma^2}\frac{\omega^2\mu^2/c^2 - k_z^2}{\left(\omega - k_z v_{\parallel}\right)^2}\right] = 0 \tag{6.18}$$

where $k_\perp$ is the perpendicular component of the wavenumber, which allows for the propagation of the radiation away from the beam at the Čerenkov angle. This quartic equation has two real roots and a complex conjugate pair for the case $v_{\parallel} > c/\mu$. One wave moves opposite to the electrons, while the other three waves move parallel and couple the EM wave to the space-charge wave in an unstable (growing) relationship. Synchronism occurs if

$$\omega = \frac{c}{\mu}\left(k_z^2 + k_\perp^2\right)^{1/2} \tag{6.19}$$

which is just the condition for radiation at the Čerenkov angle. Taking an approximation to Eq. (6.18) and assuming ω_p is small, one obtains the exponential spatial growth coefficient well above threshold:

$$\frac{\sqrt{3}}{2}\left(\frac{\omega_p^2\omega}{2\mu^2\gamma^2}\right)^{1/3}\frac{\left(1 - 1/\beta^2\mu^2\right)^{1/3}}{c\beta} \tag{6.20}$$

Note that the gain increases as $\omega^{1/3}$ and decreases for large μ or γ. Growth will also occur at angles other than θ_c. The important point is that an intense beam will provide the internal feedback which transforms the Čerenkov process into a powerful source of radiation (induced emission). The device can operate as an amplifier in the traveling-wave mode, or as an oscillator if external feedback is provided. If the electron beam propagates in a metallic drift tube completely filled with dielectric, then $k_\perp \approx 2.4/R$ and $\omega \approx k_\perp c(\beta^2\mu^2 - 1)^{-1/2}$.

When the electron beam propagates through a hole in the dielectric, the question of the coupling of the beam to the dielectric becomes important. If

$\omega/k_z c > 1$ (fast wave), then the field in the hole $\propto J_0(k_\perp r)$ and peaks in the center, where the beam is situated. However, if $\omega/k_z c < 1$ (slow wave), the radial dependence is $\propto I_0(qr)$, where $q^2 = k_z^2 - (\omega/c)^2$; the field is a minimum at $r = 0$ and the coupling is decreased. Yet, near synchronism ($\omega = vk_z$), we have $q \sim \omega/c\beta\gamma$, and so if relativistic beams are used, the field drops off slowly compared with the hole dimension. The gain is reduced by a factor $\exp(-k_z \Delta d/\gamma^3)$, where Δd is the radial separation between the beam and the dielectric. Charging up of the dielectric from the nearby beam may contribute to degradation of beam quality.

The electron beam must be cold, otherwise the space charge wave suffers Landau damping. Nevertheless, the high gain of the Čerenkov interaction permits the operation to be continued into the Compton regime [196]. The efficiency is estimated to be $\sim (\gamma + 1)(\beta\gamma)^2\lambda/L_\mu$, where L_μ is the length of the dielectric resonator.

The Čerenkov and FEL processes can be combined by the addition of an undulator. The resonance relationship is modified by the slowing of the EM wave, so that c is replaced by c/μ [see eq 2.57]. Then

$$f_s \approx \frac{v_\parallel/l_0}{|1 - \beta_\parallel \mu|}, \qquad \text{or} \quad (\mu - 1) + \frac{\lambda_s}{l_0} \approx \frac{1 + a_\omega^2}{2\gamma^2} \qquad (6.21)$$

An advantage in using the dielectric is that the beam energy required to reach short wavelength can be reduced. Radiation will propagate axially if the pitch angle of the electron helix is the Čerenkov angle [particular $B_\perp$ —see Eq (5.17)].

The Čerenkov FEL process has also been studied using short-wavelength radiation and a high-energy (50–100 MV) electron beam [61]. A gas cell, containing either H_2 or CH_4 at pressure 1.3 or 0.45 atm respectively, provided the dielectric, through which the electrons passed at the Čerenkov angle (≈ 18 mrad) with respect to an intense optical laser beam (1.06 μm, 30 MW). Electron modulation in energy, caused by the *inverse* Čerenkov interaction, was observed to depend upon the dielectric coefficient of the gases in a predictable way. (The inverse Čerenkov effect involves the transfer of momentum from the light beam to the electron beam oriented at the Čerenkov angle.) The electron bunching maximizes downstream from the interaction zone, and at this location the beam should be capable of amplifying spontaneous Čerenkov radiation emitted locally. When the gas pressure was adjusted properly, twice enhanced emission of Čerenkov light was observed at the second harmonic, 0.53 μm.

Smith-Purcell FEL

The dielectric is not the only slow-wave structure. An EM wave propagating near a metallic wall on which periodic grooves are machined (period $= l_0$)

will have its phase velocity modified according to a Bloch-Floquet relationship, such that slow-wave coupling to the beam space-charge wave becomes possible. A device described by Nation [142] used an intense beam, a rippled wall, and a backward-wave (viz., ω/k and $\partial\omega/\partial k$ have opposite sign) feedback interaction to generate powerful microwaves. More recently, the Smith-Purcell effect [179] has inspired the invention of the *orotron*, a slow-wave device which operates in the millimeter spectral range.

To use the Smith-Purcell effect in a way to generate useful amounts of power, synchronism is obtained by matching the axial wavenumber of the beam space-charge wave, $\omega/v_{\parallel}$, with the first-order spatial harmonic of the slowed EM wave, $k_z + k_0$ ($k_0 = 2\pi/l_0$). The radiation will be emitted at angle ϕ_0 with respect to the beam, where $\cos\phi_0 = k_z/k$; and of course, $\omega = kc$. Thus

$$\frac{\lambda}{l_0} = \frac{1}{\beta_{\parallel}} - \cos\phi_0 \tag{6.22}$$

An FEL-type relationship obtains for $\phi_0 \approx 0$, but in practice an electron beam is passed just above a grating and a reflecting wall is located over the grating to configure either a waveguide or quasi-optical resonator, reflecting waves at $\phi_0 \approx \pi/2$. Then the frequency generated is just $v_{\parallel}/l_0$.

The first Smith-Purcell device was named the orotron, and was constructed by Rusin and Bogomolov [165]; a millimeter-wave version has been recently described by Leavitt and Wortman [117]. A comparison of the Čerenkov and the Smith-Purcell FELs with the conventional FEL has been made by Gover and Sprangle [86].

7

FEL Experiments at Long Wavelength

7.1 *Introduction*

We now focus attention on FEL experimentation at relatively long wavelength, specifically in the submillimeter regime. In this area of the spectrum, the FEL competes only with the few isolated, relatively inefficient (but still very useful) molecular lasers. With its combination of tunability, power, and pulse-shape properties, it is highly likely that the FEL can make important contributions to physics by providing coherent power in this spectral region (Chapter 1).

FEL research in this domain has not been so fortunate in the availability of reliable, precision accelerators. The thrust of experimental physics has been toward higher energy, not higher-current beams, at least until comparatively recently. The technology and understanding of intense beams at rather low energy (1–5 MV) is rather specialized toward applications which have little to do with the FEL, and even the microtron was a low-current stepchild of the higher-energy accelerator until the last few years. Investment in new accelerator technology specifically for the FEL has been minimal, and the field needs to develop a battery of diagnostic methods as well. Another nagging problem is that the usual array of EM detection elements and instruments is underdeveloped in the submillimeter region, owing to the absence of good sources. The design and calibration of such equipment is a project in itself.

It comes as no surprise, then, that FEL experimental progress in the long-wavelength Raman region has been as slow as in the short-wavelength region, where projects must wait for accelerator time. Experiments have been done in the superfluorescent [single-pass noise amplifier] and oscillator configurations, neither of which is optimum for detailed comparison with theory—which has concentrated on the amplifier model. In the example of Raman oscillators, simple criteria have been found for regeneration, yet the single-pass gain has not been measured accurately, because of uncertainties about cavity losses, filling factor, and non-plane-wave modes. The main advance has been in the use of spectroscopic diagnostics, which have definitively established the Raman mechanism and have provided information about the dependence on pump amplitude or quiver velocity. High power and efficiency have been demonstrated.

In this chapter, we shall summarize what is known about the Raman FEL, considering experimental work in the oscillator, superfluorescent, and amplifier configurations. In this device, a strong guiding magnetic field is essential to provide an equilibrium for the electron beam in the presence of appreciable space-charge forces. We conclude with a discussion of novel collective FELs based upon the more complicated orbits which result when the motion of an electron has an important component perpendicular to the guiding field, even in the absence of an undulator: we referred to this as the "cyclotron-undulator" FEL in Chapter 6. Thus, in this chapter we are summarizing experimental information for those FELs where the guiding magnetic field has a crucial effect on the operation.

7.2 *Accelerators*

Most Raman FEL experiments done thus far have used the pulsed-transmission-line device, but more recently the pulsed induction linac has been used. Both are single-shot devices insofar as they are used in research, but are capable of a repetition rate ~ 100/sec if need arises. The pulse duration of the former is in the range 30–300 nsec, and the latter is capable of 2-μsec pulses, while both yield currents in the kiloampere range or higher. If the upgraded microtron (see Section 8.2) can achieve current density in the electron beam of perhaps 100 A/cm^2 or higher, this will put its associated FEL in the borderline region of the Raman effect.

Pulse-Line Accelerator

In the pulse-line device, a Marx generator drives a pulse-forming transmission line, which provides a square pulse output to a matched load, the accelerator diode. Use of a Marx generator by itself is usually unsatisfactory (unless it drives a special transformer) in view of the long risetime of the circuit. In the high-power diode, emission of plasma from anode bombardment will cause appreciable changes in impedance in several hundred nanoseconds, and this means that the risetime of the accelerator pulse must be short. Therefore, the Marx generator is used to charge a transmission line, which is isolated from the diode by a high-voltage switch. The line is filled with a high-dielectric-constant ($\approx$ 80) liquid such as water, so that the energy density is high and the wave speed is low. The pulse duration is twice the length of the line. Geometrical limitations relating to dielectric failure set the line impedence in the range of 20 Ω, and this must of course be the load (diode) impedence as well. Hence, appreciable current—more than the FEL needs—is provided once $V \geq 1$ MV. Some current can be diverted to a parallel resistor load, and the beam can be apertured (Fig. 7.1) so that $I_B/(I_D + I_R) \ll 1$. This procedure merely adapts rather inefficiently the device for FEL applications. Another technique to reduce the effect of too high current is to form the beam as a thin cylindrical shell rather close to the grounded wall of the drift tube.

The full diode voltage is usually applied directly across a gap in machines where $V < 3$ MV. Graphite is a suitable material for both anode and cathode. In such accelerators, flashover at the diode insulator places a practical limit $\approx$ 3 MV on the voltage. Electron bombardment of the anode surface releases a plasma which expands across the gap at the rate of a few cm/μsec: this causes the diode impedance to change, and for a matched system the accelerator voltage will vary as $\Delta V_d/V_d \propto \Delta d/d$, where d is the cathode-anode gap [163]. It has been observed [178] that the emittance of such diodes also increases as closure advances, but they find that $v_\perp/v_\parallel$ can be < 0.05 for a pulse time as long as 150 nsec using a cold cathode. A switch is introduced just before the diode to prevent a lower-voltage prepulse from appearing on the cathode.

Shown in Fig. 7.1 is a pulse-line accelerator designed by Physics International and installed at Columbia University. Special attention to the design resulted in an accelerator voltage pulse into a typical FEL diode which was

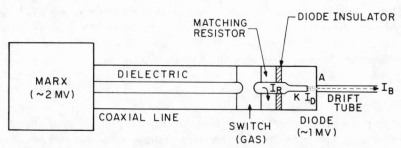

Figure 7.1 Schematic of pulse-line accelerator at Columbia University.

flat to within 2% for over 100 nsec (see Fig. 7.13a); an eccentrically mounted inner conductor in the transmission line permitted compensation for the change in diode impedence during the pulse. Note that only $\frac{1}{2}$ of the Marx voltage appears at the diode. Use of a different pulse line, referred to as a Blumlein (after its inventor), will permit the diode voltage to equal or exceed the Marx voltage.

Induction Linac

A high-current accelerator which is more complicated but perhaps better suited to the FEL is the induction linac (Fig. 7.2). In an induction linac, a number of pulsed modules are arranged in series, linked by the electron beam. The electric field at the accelerating gap is produced inductively, so that the separate voltages from the modules will add only at the center of the device; large electrode potentials are therefore avoided.

The ETA device at Livermore has produced 5 MV and 10 kA, and a scaled-up version is to operate at 50 MV and 10 kA, providing 30-nsec pulses. A long-pulse version was built at NBS [118] and is now in operation at NRL in an FEL experiment. The injector portion includes an electron gun (either hot or cold cathode) and several control electrodes operating at 400 kV altogether, and 0.8 kA. Electrons are accelerated in the gun through a series of 12 annular electrodes. A graphite brush cathode has given satisfactory service.

Focusing coils transport the beam to an array of induction accelerator modules, which develope about 200 kV across each accelerating gap. In the NRL effort, about 200 A of beam current was transported to the undulator; the temporal variation of beam energy was < 3% over 1.6 μsec, measured at the injector. However, the emittance of the transported beam in the vicinity of the undulator was ~ 1 cm-rad when a guiding field was used to transport the beam. Considerable improvement resulted when the guiding field was replaced entirely with a set of focusing magnets.

7.3   The Raman FEL

Prior to the first Raman FEL [132], a series of experiments had been done which uncovered some of the important physics behind the device.

Early experimentation with intense, megavolt electron beams using a periodic rippled field from an iron ring undulator showed that large amounts of microwave radiation could be produced [80]. Further experiments by Granatstein et al. [87] showed that the electron-cyclotron-maser mechanism was at work, and this was confirmed by Talmadge et al. [189],

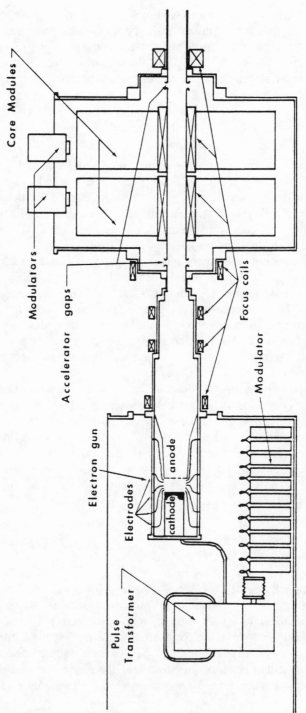

Figure 7.2 Schematic of long-pulse induction linac accelerator at NRL [163]. © 1983 Academic Press.

who demonstrated that a short magnetoresonant undulator in fact produced a beam with appreciable transverse electron velocity. In another experiment, appreciable radiation at 400 μm was detected; this was interpreted as the result of stimulated backscattering of microwaves on the cold portion of the beam prior to entry into the undulator [88]. The undulator would transform part of the electron streaming energy irreversibly into transverse motion if the period was chosen to satisfy the magnetoresonance condition; following this, the cyclotron maser instability generates a few hundred megawatts at the cyclotron frequency; this intense microwave beam reflects at the end of the device, and acts as an electromagnetic pump on the upstream cold beam to produce radiation at frequency $\approx 4\gamma_{\parallel}^2(\Omega_0/\gamma)$.

In a different approach, the undulator was extended for perhaps 25–50 periods, its amplitude was reduced to perhaps 10% of the guiding field, and no magnetoresonance was desired. Then, the undulator produces only a reversible transverse quiver motion of the electrons. A spectroscopic study of superradiant emissions from such an arrangement was undertaken by Efthimion and Schlesinger [64], who identified the coupling of the unstable slow space charge and cyclotron waves to the waveguide modes of the drift tube, at wavelength ≈ 0.5–2 cm. This work was extended to wavelengths ≈ 1–3 mm by Gilgenbach et al. [84] after substantial power was found by Marshall et al. [131]. The dependence on magnetic field was found (i.e., shorter wavelength for smaller guiding field for the cyclotron mode, and no dependence on B_0 for the space-charge mode), and the spectroscopic data supported the following scattering relationships:

$$\omega_s \approx \begin{cases} 2\gamma_{\parallel}^2\left(k_0 v_{\parallel} - \omega_p/\gamma\right) & \text{(space-charge idler)} \\ 2\gamma_{\parallel}^2\left(k_0 v_{\parallel} - \Omega_0/\gamma\right) & \text{(cyclotron idler)} \end{cases} \tag{7.1}$$

Using a pulsed undulator, Gilgenbach determined the dependence of scattered power upon undulator amplitude and located a threshold pump condition, required by theory to couple the three waves in an unstable relationship. In these experiments, the excitation of the cyclotron mode was caused by the electron beam entering an undulator in which the field amplitude was ungraded, whereupon a surface perturbation was launched upon the beam. Subsequent work [89] showed that this transient can be the cause of appreciable noise radiation on the beam, which is amplified by the FEL mechanism in a single-pass superfluorescent configuration. Given that megawatt-level power was produced, but that the FEL power gain was only of order 100, appreciable power was being supplied by beam noise at the undulator input. In later experiments (e.g., [132] or [30]) the undulator field was properly graduated and the superfluorescent signal fell to very low amplitude except during the initial transient as the electron beam is switched on. Under these circumstances, optical feedback (oscillator configuration) was necessary to restore a powerful output signal in the submillimeter spectrum. Gilgenbach found that the superfluorescent linewidth, $\Delta\lambda/\lambda \approx$ 10%, included inhomogeneous and homogeneous broadening: the latter depended linearly on the undulator field, as indicated by Eq. (4.34), while the former—obtained in the limit $B_\perp \to 0$—was consistent with $(\delta\gamma/\gamma)_{\parallel} \approx$ 3%.

In parallel with some of this work, the TRW group undertook the investigation of a simple ubitron-type FEL, which produced output in the centimeter region of the spectrum. Several waveguide modes were analyzed. One novel feature of this work was experimentation involving a signal which was injected at the FEL input to pre-bunch the electrons [35]. A group at Tomsk has also studied radiation emitted by an intense electron beam interacting with an undulator [57].

Oscillator

The Raman FEL was constructed as a joint Columbia-NRL effort, and a successful oscillator was reported in 1978 [132]; more data were taken at Columbia by Birkett et al. [30] using the same configuration. Referring to Fig. 7.3, a cylindrical shell electron beam, $R = 20$ mm and $\Delta R = 1$ mm, produced by field emission at the cold cathode, was expanded out near the walls of the drift tube by following an adiabatic taper of the guiding field. Due to high-beam current, the beam cannot be bent around the mirrors as it

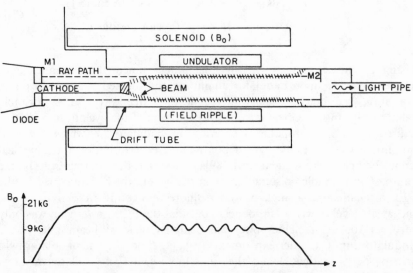

Figure 7.3 The Raman FEL oscillator of McDermott et al. [132]. © 1978 APS.

could in the two-wave FEL (Fig. 1.1). Low-γ, high-intensity current streams require special magnetic techniques to provide a stable equilibrium in such complex geometry [20, 103]. The mirrors were planar, made of highly polished stainless steel; high reflectivity is not important, because the Raman FEL gain is large and the diffraction losses are substantial.

The oscillator was to operate in the quasi-optical regime to reduce diffraction, and a wavelength $\sim \frac{1}{2}$ mm was chosen at $V_d = 1$ MV: this required a short-period undulator ($l_0 = 0.83$ cm) and location of the electron beam rather close to the undulator windings (the penalty associated with this location is disproportionate velocity shear and high harmonic content in the electron quiver motion). Nevertheless, $B_\perp$ could not exceed ≈ 500 G, or $\beta_\perp \sim 1\%$, because of the electrical and mechanical limitations on the undulator, and so this oscillator was restricted to the very-weak-pump regime. The beam entered the undulator in such a way that the undulator field amplitude was *graded*, i.e., it increased gradually along the axis. The location of the beam will excite an EM mode in the cavity with a field minimum on axis, where the cathode and the resonator coupling hole extract power from the Fabry-Perot resonator.

The major source of velocity shear is the diode geometry. Since $v_{\perp 0} \propto B_0^{-1}$, a high magnetic field was imposed at the cathode. The electric field at the cathode is largely transverse and is $\sim V_d/d$, where d is the annular gap. Increasing B_{0k} reduces the electron gyroradius and therefore the potential drop that the electron samples in the radial electric field. The spread of electron energies across the gyro-orbit is $\sim 2r_c/d$; hence

$$\left(\frac{\delta\omega}{\omega} \right)_\parallel \sim \frac{2}{d} \left(\frac{\gamma mc^2}{eB_{0k}} \right) \left(\frac{v_{\perp 0}}{c} \approx \frac{E_\perp}{\sqrt{2}\, B_{0k}} \right) \tag{7.2}$$

which scales as V_d^2/B_{0k}^2. The factor $2^{-1/2}$ in the $v_{\perp 0}/c$ term above arises from the reduction of the transverse motion of the electron as it moves into the weaker magnetic field downstream. The motion is determined by the adiabatic invariant

$$\frac{\gamma m v_{\perp 0}^2(z)}{B_0(z)} \tag{7.3}$$

and as $B_0(z)$ weakens, the transverse electron motion is partly converted into parallel motion. Other factors tending to cool the beam are that as the electrons drift toward the wall, they fall through additional potential in the radial space-charge field, while large-gyroradius electrons are scraped off on the wall. Taking $\gamma \approx 3$, $d = 12$ mm, $B_{0k} = 20$ kG, and $B_0 = 10$ kG, we have $(\delta\omega/\omega)_\parallel \approx 4\%$, which places the experiment on the Raman Compton borderline. Referring to Eq. (6.7) and taking $L = 50$ cm, $(\delta\gamma/\gamma)_\parallel$ can be at most 2%—beyond this point the gain will drop.

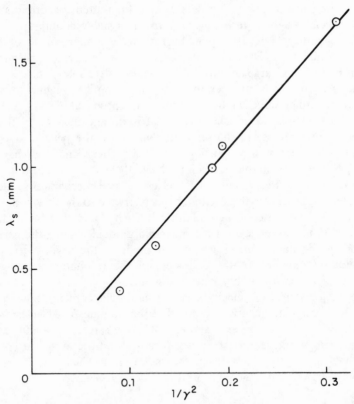

Figure 7.4 Data from experiments confirming the FEL wavelength relationship. Undulator period $l_0 = 8.3$ mm [30]. © 1981 IEEE.

Parametric interaction with the space-charge idler only was obtained, and spectroscopic analysis of the data (Fig. 7.4) showed that the FEL relationship was satisfied. The wavelength did not depend on the guiding magnetic field. The dependence of FEL power upon the undulator pump field is shown in Fig. 7.5, where it can be seen that at $B_\perp \approx 200$ G the system passes a threshold and begins to oscillate.

A typical FEL pulse shape is shown in Fig. 7.6, together with an analysis [30] for the coefficient of the exponential signal gain per pass, ΓL. Note that oscillation is sustained only during the flat portion of the accelerator pulse. The "tail" at the end of the FEL radiation pulse represents the decay of radiation contained in the resonator, and gives $\nu_L^{-1} \approx 10$–20 nsec for the radiation decay time ($\approx 50\%$ loss per pass). An independent estimate can be made for the threshold value of the gain, $(\Gamma L)_T$, using Eq. (4.44). Assume that the space-charge wave is becoming Landau-damped, but is still collec-

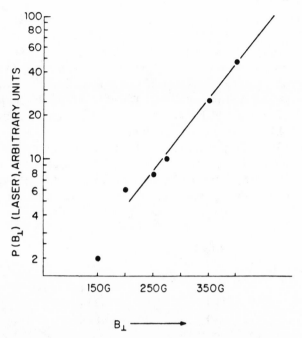

Figure 7.5 Dependence of FEL power on pump amplitude. Oscillation threshold is at $B_\perp \approx 200$ G [30]. © 1981 IEEE.

tive, and take the idler-wave damping time to be of order $10\omega_p^{-1} \sim 10^{-9}$ sec; then $(\Gamma L)_T \approx \frac{1}{2}$, about one-half to one-third the amount computed from analysis of Fig. 7.6. Having checked the two methods for estimating ΓL from the data, we settle on $\Gamma \sim 10^{-2}$ cm^{-1} and compare this with the prediction from Raman theory [Eq. (4.33)], using experimental values for $B_\perp$, l_0, and ω_p:

$$\Gamma = \frac{eB_\perp}{mc}\left(\frac{\omega_p l_0 f}{8\pi\gamma c^3}\right)^{1/2} \qquad (7.4)$$

Within a factor of 2, taking the filling factor $f \approx \frac{1}{4}$, this gives a reasonable prediction for the observed gain. There are two compensating factors which should be noted, however. The first is the effect of the guiding magnetic field, which will enhance the growth roughly a factor of 2 for this choice of parameters ([74]; see Fig. 6.5). On the other hand, the beam energy spread is as much as 4%, and we can expect a substantial decrease in gain from this source. Therefore, only approximate agreement of the measured gain with Eq. (7.4) is to be expected.

What power should we expect from this oscillator? The maximum efficiency would be $\approx \omega_p / \gamma k_0 c \approx 5\%$, but we are entitled to only a fraction g/ω_p of that, giving $\eta \approx 0.05\%$. The beam power is about 10^{10} W [10^4 A $\times 10^6$ V], and so we would expect ≈ 5 MW, which is approximately the actual power level in the resonator (the power loss, divided between the two ends, gives the observed power $\sim 10^6$ W emitted from the coupling hole). Much larger output and higher efficiency could be obtained in principle by pumping the FEL harder, up to the limit where $g \sim \omega_p$. This would also have the beneficial effect of stabilizing the oscillator output against fluctuations in accelerator voltage. In an oscillator, the EM energy is retained in the resonator, and therefore the frequency is fixed; if this EM energy is to grow, the electron energy also must be fixed within the rather narrow

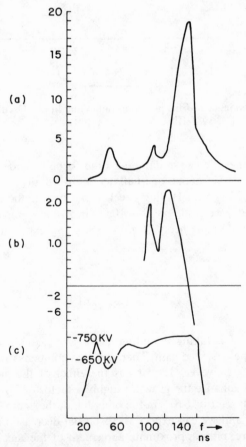

Figure 7.6 Calculation of ΓL for $\gamma = 2.3$, $\lambda_s = 1.0$ mm: (a) power amplitude; (b) ΓL; (c) diode voltage [30]. © 1981 IEEE.

bounds defined by the spectrum of unstable wavenumbers (see Fig. 4.1). Increasing g to of order ω_p, still within the bounds of Raman backscattering, will accommodate the accelerator voltage ripple present in pulse-line apparatus of the type used [$\Delta V_d/V_d \approx 2\%$]. However, it was not possible to increase the pump amplitude to this extent, and therefore it was possible neither to stabilize the power of the oscillator to one level (shot-to-shot variations were common), nor to drive the oscillator to the limit of its theoretical efficiency. To cure these problems, the undulator period and pump field should be increased and the electron beam located on the axis where the undulator velocity shear is lowest. But, before jumping ahead to a new design, we shall first comment upon some other interesting features in these first Raman-oscillator experiments.

A small fraction of the EM energy stored in the resonator was coupled out via a hole in the anode mirror. There were also substantial diffraction losses, which are desirable and affordable, the Raman gain being large. The high loss selects out the dominant EM mode of the pipe, which is represented by the Bessel function $J_1(k_\perp r)$. The higher-order radial modes, representing slightly off-axis propagation, are unstable [114], but will not regenerate, since their losses in this "open" resonator are very large [30]. The EM waves are the guided waves of a cylindrical pipe, which convert to free-space modes in the region between the end of the pipe and the mirror, and vice versa. In the region of the undulator, each waveguide mode would travel with a slightly different phase velocity due to the finite ratio of wavelength to drift-tube radius. The spread of axial wavenumbers is, from the waveguide dispersion relation,

$$\frac{\delta k_z}{k_z} \approx \frac{1}{2}\left(\frac{\lambda_s}{R}\right)^2 \sim 10^{-3} \tag{7.5}$$

Taking $\lambda_s \approx \frac{1}{2}$ mm and $R = 25$ mm, the spread of EM wave phases due to excitation of nonaxially propagating waves is $\delta k_z L \approx k_z \cdot L/1000 \approx 2\pi$: the various waveguide modes would be interfering with each other. It is better to select only the dominant mode, by "opening" the ends of the resonator. If the Rayleigh range is taken to be the undulator length, then $R \approx \sqrt{\lambda_s L/\pi}$ and the diffraction angle is $\theta_D \approx 1/\gamma N^{1/2}$. The condition that absorption via the anti-Stokes mode shall not occur, $\theta_{0A} > \theta_D$ [Eq. (4.42)], is satisfied if the collective FEL condition [Eq. (2.60)] is met.

Using a Fabry-Perot interferometer, narrowing ($\delta\lambda/\lambda \approx 2\%$) of the oscillator spectrum was observed by both McDermott (Fig. 7.7) and Birkett as the oscillation became established. This is a marked improvement over the expected inhomogeneous line width $(\delta\lambda/\lambda)_{inhom} \approx 2(\delta\gamma/\gamma)_{\parallel} \approx 10\%$ (the observed linewidth of the radiation emitted from a similar super-fluorescent FEL is of this magnitude [84]). The FEL inhomogeneous line width is still large compared with the resonator line width ($\sim \omega_s/Q$), and

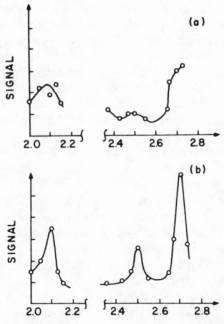

Figure 7.7 Fabry-Perot interferometer fringes, showing how the fringe visibility improves: (a) early in FEL pulse, signal growing; (b) late in FEL pulse. Horizontal axis: etalon separation x; $\Delta\lambda/\lambda = \Delta x/x$. © 1980 Academic Press.

many axial resonator modes, spaced apart by $c/2L$, should be excited. Oscillating cavity modes are those clustered about the frequency for which the FEL gain exceeds threshold, and these form a narrowed subset of the entire gain spectrum at low gain.

Superfluorescent Amplifier

A redesign of the FEL was undertaken next by the group at NRL [151] (Fig. 7.8). They chose to locate the electron beam on the axis of a longer period ($l_0 = 3$ cm) undulator which would permit strong-pump operation at high power and efficiency. Moving the beam to the axial location required a reduction in beam current by roughly a factor of ten (~ 1 kA), which was done by aperturing the beam in the diode. An extensive program of numerical modeling showed how this was to be done without incurring large velocity spread or surface rippling on the beam. To obtain a strong pump wave, the size of the drift tube was reduced to 11-mm diameter. To avoid the use of mirrors, it was decided to operate in the vicinity of magnetoresonance, where the gain of the system would be so large that noise power at the FEL input would grow to many megawatts at the output, 60 cm downstream (gain ≈ 1.2 dB/cm was observed). The large electron quiver

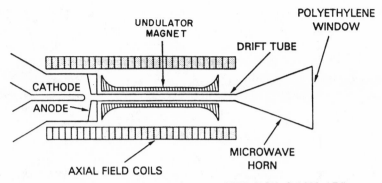

Figure 7.8 An FEL experiment at NRL [151]. © 1982 APS.

motion in the strong undulator field ($\beta_\perp \approx 0.3$) accounts for the lion's share of the inhomogeneous broadening: $(\delta\gamma/\gamma)_{\parallel} \approx 5\%$, from Eq. (5.29). This is however still within the Raman criterion, as the scattered wavelength is ≈ 4 mm.

The NRL group found that high power and efficiency could be obtained using their Raman FEL in the high-gain superfluorescent configuration. The latest report is that about 75 MW was obtained at roughly 4 mm, which represents an efficiency $\approx 7\%$. By examining the spectrum (Fig. 7.9), two

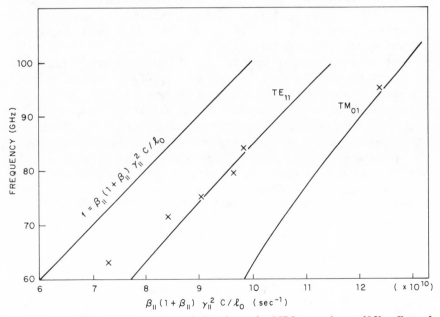

Figure 7.9 Frequency of FEL emission from the NRL experiment [85]: effect of waveguide cutoff. © 1983 AIP.

different radiating waveguide modes were observed, the TE_{11} and TM_{01} cylindrical modes of the drift tube. The proximity of the scattered wavelength to the cutoff wavelength of the drift tube requires the use of the guided-wave EM dispersion relation, $\omega^2 = \omega_c^2 + k_z^2 c^2$, to solve for the FEL frequency (limit of $\omega_p \to 0$):

$$\omega_s = \gamma_{\parallel}^2 k_0 v_{\parallel} \left[1 \pm \left(\beta_{\parallel}^2 - \frac{\omega_c^2}{\gamma_{\parallel}^2 k_0^2 v_{\parallel}^2} \right)^{1/2} \right] \tag{7.6}$$

where each mode has a different ω_c. As a result, the FEL frequency is reduced below the free-space result. Choosing $V_d = 1.25$ MV and taking into account that $\gamma_{\parallel}$ is reduced by the large quiver motion, the scattered wavelengths expected (solid line, Fig. 7.9) correspond to the observed values.

When operating in the vicinity of magnetoresonance, the gain formula (7.4) must be modified by a multiplying factor $\Phi^{1/4}$, where

$$\Phi \equiv 1 - \frac{\Omega_0 \gamma_{\parallel}^2 v_{\perp}^2}{\left(v_{\parallel}^2 + v_{\perp}^2 \right) \Omega_0 - \gamma k_0 v_{\parallel}^3} \tag{7.7}$$

which comes from the ponderomotive potential. For group I orbits (Fig. 5.2), $\Phi > 0$ and the gain is enhanced; however, $\Phi < 0$ for the group II orbits near magnetoresonance [74]. Under such conditions, the plasma space-charge wave identity is lost, and there is no downshift in frequency by the factor ω_p/γ as in Eq. (7.1). The group II orbit gain remains high, ≈ 2 dB/cm under experimental conditions.

Operation near magnetoresonance with high electron quiver velocity in the undulator opens the possibility of a new type of efficiency enhancement [75], wherein the axial magnetic field is changed adiabatically along the undulator. The average axial electron velocity and hence the phase relationship between the electrons and the optical wave remains fixed; however, the tapering of B_0 permits the reduction of the electron quiver kinetic energy as the radiation is amplified. The pendulum equation once again will describe this situation, with the "constant torque" term $\sin \psi_r$ now represented by a term involving $(\Phi - 1) d(\ln B_0)/dz$. Enhancement of the FEL radiation occurs for group I orbits if $B_0(z)$ decreases, as well as for group II orbits if $\Phi < 0$. The efficiency is

$$\approx \beta_{\perp}^2 L_0 \left(\frac{1 - \Phi}{\Phi} \right) \frac{d}{dz} \ln B_0$$

where L_0 is the region of tapered field. This mechanism accounts for the doubling of FEL power observed by the NRL group when the uniform guide magnetic field was replaced with an adiabatic gradient.

The data show that the spectrum is strongly affected by changes in the quiver velocity at constant B_0 [which is expected, according to Eq. (2.10)],

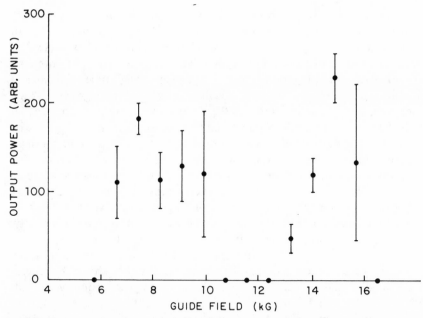

Figure 7.10 Output power of NRL FEL near magnetoresonance ($B_0 \approx 11$ kG) [100]. © 1983 IEEE.

but the spectrum is unchanged if $\beta_\perp$ is fixed while B_0 is varied. This is strong evidence that the FEL radiation, rather than cyclotron emission, is being produced. As magnetoresonance is approached, the radiation falls abruptly (Fig. 7.10), an effect also observed by Birkett and Marshall [31]. This can be explained by several effects: (a) a drop in gain caused by enhancement of inhomogeneous broadening; (b) shift of the radiation below the waveguide cutoff owing to the large quiver velocity at magnetoresonance; (c) reduction of beam current owing to the scrapeoff of electrons having large helical orbits; (d) excitation of a convective microwave mode [123].

Converting the experiment to an oscillator would require a reflector at the undulator input. However, a metallic element cannot be introduced as a discontinuity in the drift tube without perturbing the electron-beam equilibrium. For the long wavelength generated in this experiment, mirrors are not useful. An ingenious substitute has been proposed by Efthimion: it consists of a thin, tapered cylindrical wedge of dielectric positioned near the wall of the drift tube. The dominant cavity waveguide mode is coupled gradually onto the wedge via the dielectric surface wave. At the blunt end of the wedge, the surface wave is reflected, whereupon it converts back into the cavity mode as it retraces its path.

In a Raman FEL, there is competition between the convective FEL instability at high frequency, and another, *absolute* instability in the microwave region. The latter corresponds to a wave propagating counter to the electron flow, and, according to the treatment in Chapter 2, the wavelength of this mode is $\approx 2l_0$. This instability has been studied by Kwan [114] as part of a program of FEL numerical simulation. It was found that although the FEL signal initially grows more rapidly, given enough time or distance it will saturate; the slower-growing absolute instability then grows to high amplitude, whereupon it dominates the electron-beam physics and turns off the higher-frequency mode. Kwan noted that the absolute instability may be suppressed by choosing the dimension of the waveguide so that the low-frequency mode is below cutoff. Liewer et al. [119] have found criteria for operation of the FEL in a regime which discriminates against the absolute instability. There are three characteristic lengths involved: L, Γ^{-1}, and L_{crit}, which are respectively the undulator length, the growth length of the FEL signal, and the critical length for onset of the absolute instability. Numerical simulation shows that the FEL mode will saturate in ≈ 7 multiples of Γ^{-1}, and so an FEL should have $L = 7\Gamma^{-1}$ to saturate the output; it is furthermore required that $L_{\text{crit}} < L$. Fig. 7.11 shows the criteria to be satisfied. The FEL experiments described in this section would fall into the stable operating regime for $\gamma > 2.8$, since $ck_0/\omega_p \approx \pi$.

In recent experiments with the long-pulse (2 μsec), 700-kV linac at NRL, Pasour et al. [152] have identified Raman FEL emission at 30 GHz. A cold-cathode carbon brush cathode emitted the electron beam. Focusing coils, a guiding field, and helical undulator were used to transport the beam,

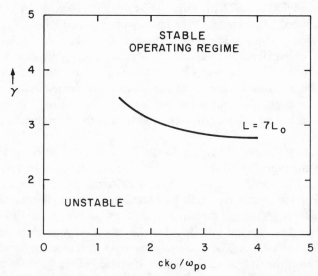

Figure 7.11 Criteria for stability or instability of the absolute forward-scattered mode [119]. © 1981 APS.

which carried a current of 200 A with emittance ~ 1 rad-cm. The latter implies a substantial $v_{\perp 0}/v_{\parallel}$, which may be responsible for the substantial cyclotron emission described in their first report (a similar interpretation was made for data reported by Shefer and Bekefi [172]); in addition there was FEL radiation at low level. The difficulty here is that the beam is created in zero guiding field, and when the latter is increased, there is a tendency to "spin up" whatever transverse motion the electrons acquired in the electron gun [see Eq. (7.3)]. When the guiding magnetic field was removed—in which case focusing elements alone directed the beam into the helical undulator—the cyclotron emission disappeared, leaving a much-enhanced FEL line. To date, the efficiency obtained has been good, about 1.5% with a uniform undulator, representing about 2 MW of power. They have also observed the low-frequency forward-scattered mode at 100-kW power level.

Bratman [38] has proposed a type of two-stage FEL device, in which a periodic rippled wall of the drift tube is used to generate a strong EM backward wave via the electron-beam—slow-EM-wave interaction. In an experiment at NRL, Carmel et al. [42] have reported that about 300–500 MW can be generated at $f_0 = 12.5$ GHz in the backward-wave interaction, using an electron-beam current and energy similar to that driving the NRL Raman FEL. This EM wave is equivalent to an undulator with period $l_0 = c/f_0(1 + v_{\parallel}/v_\rho) \approx 1.7$ cm and amplitude $B_{\perp} \approx 200$ G (the microwave phase velocity v_ρ was 7×10^{10} cm/sec against the electron flow). This powerful wave stimulates short-wavelength Raman FEL emission produced either at the same location as the slow wave structure, or possibly further upstream. About 0.35 MW was obtained at $\lambda_s = 1.5$ mm in a single-pass configuration. In addition, a forward-scattered FEL mode, of comparable power, appeared at 8 GHz and was identified as an absolute instability.

Amplifier

With regard to Raman traveling-wave FELs, some data at $\lambda_s \approx 3$ cm has been reported by the TRW group [35], and another project is underway at Livermore [199], where amplifier measurements at ~ 1 cm are being made with an induction linac machine ($\gamma = 7.4$). Amplification of a high power (20 KW) coherent signal has been observed at LLL, obtaining good efficiency in the strong pump regime. The NRL experiment has reported Raman FEL traveling-wave gain as high as 50 dB, obtaining many megawatts of coherent power at 35 GHz. Preliminary indications are that the power coupling loss is roughly as given in Chapter 6, in the range -12 to -15 dB.

At Columbia University, a Raman FEL amplifier has operated at 1.2 mm, where coherent radiation is provided by an isotopic $^{13}CH_3F$ laser. This molecular laser is pumped by a 9.6 μm CO_2 laser; a pumping power ≈ 1

MW yields about 10 kW at 1.2 mm. The signal is injected into the drift tube downstream, above the undulator, in an electron-beam configuration very similar to that used in the NRL experiment just described in the preceding subsection (Fig. 7.8). In fact, the system is the very same as used for the Thomson-scattering determination of the beam parallel-velocity spread, mentioned in the last subsection of section 5.5. Accordingly, energy-spread data are available as an input to the theory. The coherent wave is injected through a mesh which forms part of the drift-tube wall upstream from the undulator entrance, after which single-pass amplification results. The beam energy is $\gamma \approx 2.5$; for the undulator, $l_0 = 1.1$ cm, $B_\perp = 0.6$ kG; the total inhomogeneous broadening is $(\delta\gamma/\gamma)_\parallel \approx 1.5\%$; the filling factor $\approx \frac{2}{3}$; and B_0 is adjusted to about 10 kG so that $\beta_\perp \gtrsim 0.05$. Taking into account the filling factor and the coupling loss, a 50-period undulator with a 1-kA/cm^2 beam provides > 6-dB gain of the total power incident on the undulator; if the undulator is extended to the cold-beam limit, about 75 periods, then the gain should be about 15 dB.

7.4 Hybrid FEL

The theory of this device, discussed in the final subsection of Section 6.3, predicted the possibility of a cyclotron-maser–FEL hybrid, wherein the double-Doppler-shifted frequency behavior of the FEL could be combined with an instability promoted by the large gyroradius electron orbits. To test this concept, Grossman et al. [90] constructed a variation of the FEL resonator used in the Raman experiments, using the same drift tube, anode mirror, and electron-beam geometry (Fig. 7.12). However, the diode is designed so that electron gyromotion in the fringing transverse electric field at the diode is enhanced by compression of the guiding field, so that $\beta_\perp \approx 0.4$ and $\beta_\parallel \approx 0.8$ are obtained. A tapered undulator, $l_0 = 1.25$ cm and $L \approx 50$ cm, provides electron quiver motion up to $\beta_\perp \approx 0.1$ at the beam location.

Power was obtained (< 1 MW) at $\lambda_s \approx 1.5$ mm; some typical pulse shapes are shown in Fig. 7.13. The power is characterized by rapid growth to saturation, and an abrupt decrease at the end of the diode voltage pulse as the parametric resonance is broken, followed by an exponential decay of energy stored in the resonator. Negligible power is obtained when the feedback is removed (Fig. 7.13c) by taking away the anode mirror. Spectral analysis showed [92] the presence of harmonics, the decrease of FEL wavelength with increasing B_0, and the dependence on $\gamma_\parallel$. The short wavelength radiation was consistent with the prediction of Eq. (6.12). Compatible values of $\beta_{\perp 0} \approx 0.4$ and $\beta_\perp \approx 0.1$ are necessary for oscillation in the resonator (note that $\beta_\perp$ is about a factor of 4 large than that required

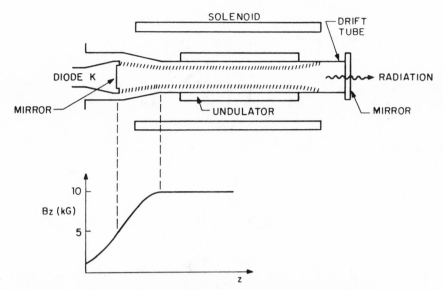

Figure 7.12 Hybrid FEL configuration [90]. © 1983 IEEE.

for the operation of the Raman FEL oscillator). In another experiment [90], Grossman et al. demonstrated that the $\beta_{\perp 0}$ threshold for the hybrid FEL instability radiation was lower than that of the Doppler-shifted cyclotron instability, where no undulator is used.

The dependence of the oscillator power upon undulator field amplitude is quite different from that for the Raman FEL, and is clearly different from that of the incoherent noise (Fig. 7.14). The coherent signal is marked by an abrupt threshold, followed by saturation, whereas the noise signal continues to increase as $B_{\perp}$ is increased. Examination of the temporal behavior of one of the harmonics as well as the microwave (2–3 cm) output suggested that a possible cause for the saturation of the dominant [$p = 1$, $q = 1$ in the expression below Eq. (6.12)] FEL fundamental could be the growth of harmonics and/or the microwave background. It is well known from studies of the intense-beam cyclotron maser (e.g., [80]) that appreciable electron gyromotion can excite powerful microwave radiation in waveguides near Ω_0/γ and its harmonics. Efficient operation of this device therefore requires special attention to the suppression of unwanted modes of oscillation.

Another potentially useful hybrid is the *crossed-field FEL* or rippled-field magnetron, proposed by Bekefi [22]. Electrons are emitted from a cold cathode into an annular space in crossed DC electric (E_0) and magnetic (B_0) fields, and undergo an $\mathbf{E}_0 \times \mathbf{B}_0$ drift at relativistic speed. In cylindrical geometry the drift is sheared, since $E_0 = E_0(r)$. The axial field B_0 must exceed the critical field required for "magnetic insulation" (i.e., zero current

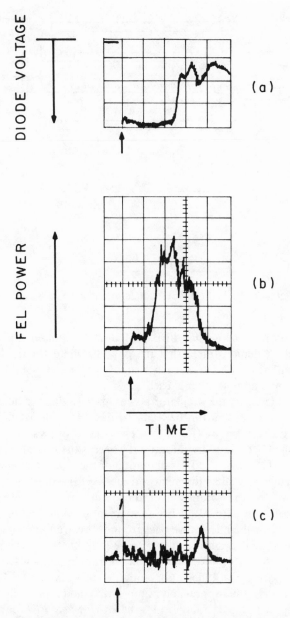

Figure 7.13 Signals from the Hybrid FEL [92]: (a) diode voltage, ≈ 800 kV; (b) radiation at 1.6 mm, both mirrors in place (20 mV/div, vertical); (c) Radiation at 1.6 mm, anode mirror removed (5 mV/div, vertical, 50 nsec/div, horizontal); arrow shows $t = 0$. © 1983 IEEE.

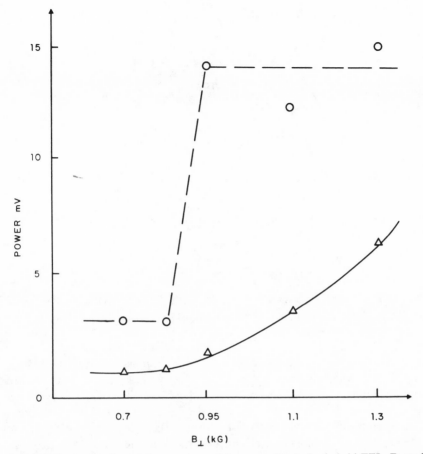

Figure 7.14 Dependence of power on undulator amplitude for hybrid FEL. Dotted line, oscillator; solid line, noise signal [92]. © 1983 IEEE.

flow to the anode) given by

$$B_{0c} = \frac{mc}{ed_0}\sqrt{\gamma^2 - 1} \tag{7.8}$$

where $d_0 = (r_c^2 - r_a^2)/2r_a$; here r_c and r_a are the cathode and anode radii (Fig. 7.15). An arrangement of samarium-cobalt magnets defines the undulator; the configuration is an oscillator, with radiation contained in the cavity and the flow of electrons closed upon itself. It does appear however that a sufficiently strong magnetic perturbation from the undulator may permit electrons to reach the anode, even if $B_0 > B_{0c}$.

If the electron stream were cold, one might expect the typical FEL radiation; however, the velocity shear may impart enough temperature to

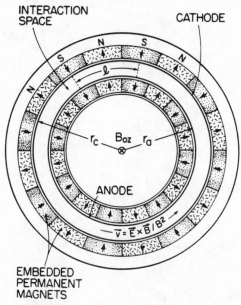

Figure 7.15 Configuration of the rippled-field magnetron [22]. © 1982 AIP.

the system such that the Raman instability is suppressed while the "hybrid" instability, which is more tolerant of velocity shear, is favored. Spectroscopic studies of the millimeter radiation [$\approx$ 300 kW] at MIT [23] show that it is strong and narrowband, and that the frequency increases with increasing B_0 in a way that appeas to fit Eq. (6.12). To reduce the velocity shear of the magnetron's radial electric field, a nearly monoenergetic, cylindrical-shell (radius R) electron beam can be injected through a magnetic cusp. This will transform most of the axial electron motion into a rotation (of an axially translating ring) at speed $R\Omega_0/\gamma$, where the gyrofrequency is computed in the axial magnetic field downstream from the cusp. The electrons remaining in the spiraling ring will find a new equilibirum, characterized by low velocity shear. An MIT-Maryland experiment using this approach has reported about 200 kW of power at short wavelength, apparantly arising from an FEL mechanism. As the low-frequency microwave output was not enhanced by the presence of the undulator, it was concluded that the undulator field does not enhance the negative-mass instability [24].

8

Two-Wave FEL
Experimentation

8.1 *Introduction*

Here we shall discuss some of the experimental progress which has been
made with FELs operated under conditions of high beam energy and low
current. Experimental information about FELs operating at short wavelength, once very limited, is now rapidly accumulating, and the field is
expanding into new territories. Limitations on resources and on accelerator
time have slowed progress: the FEL is not a tabletop device, and extensive
support is necessary before the system will oscillate. However, by 1983 an
important corner had been turned, and there are now several quite different
FELs in operation or nearing completion. Visible radiation has been produced for the first time, both from the storage ring, and on a higher FEL
harmonic using an rf linac. Experimentation with nonuniform undulators
has solidly established the utility of these devices in enhancing FEL gain
and/or efficiency.

   The FEL can now be regarded as a field for scientific as well as
promising technological work. One indication of the field's maturity is that
when the accelerator performance meets certain standards, the FEL operates as expected. High-quality optical and electron-beam diagnostics have
removed uncertainties concerning the experimental variables. The latter
appear to be properly understood theoretically, as the data will show. There
still may be barriers for future progress: some are technical (e.g., mirrors),

while others (e.g., the sideband instability) may represent inherent limitations.

The operation of an FEL is largely a category of accelerator physics, about which there is a vast literature as well as a standardized learning procedure. There are exacting requirements on beam preparation, injection, and alignment that demand both experimental skill and sophisticated instrumentation. Cavity-resonator design principles are treated in standard laser texts, e.g., [14], and will not be repeated here. Our approach in this chapter will be to gather the experimental information and present examples which pertain to the basic FEL theory already discussed. It is important to remember that testing of the FEL under experimental conditions is just beginning; to a large degree its limitations depend on technological problems, and so what appears to be impossible today may suddenly become routine tomorrow.

We begin with a brief discussion of the various accelerators which are in use. The first FEL results, those of the Stanford group (using a uniform undulator), are discussed next, and form a quantitative body of information on which to base our understanding of how the laser actually performs. There follow two sections on comparatively new FEL systems—the storage ring and the two-stage device. Nonuniform-undulator research is now a well-established technical area, and some of the research data are summarized in Section 8.6. We conclude with a discussion of the microtron submillimeter FEL, which could have been described in Chapter 7, but whose accelerator is very similar to the type under consideration here.

8.2 Accelerators

rf Linac

The rf linac has been the preferred electron-beam accelerator for the short-wavelength FEL. It provides a pulsed (see Fig. 1.6) electron source which is very reproducible and exceptionally stable. Usually the beam has good emittance; if not, it can be upgraded in a systematic way. The rf linac is driven by a pulsed source of high-power microwave radiation injected into a disk-loaded waveguide, arranged in a line, through which the beam is passed. The rf is pulsed at several hertz, for several microseconds in each pulse. During this "on" time, the acceleration occurs near the peak of the rf field. The current output of the accelerator is a sequence of micropulses, each a few picoseconds long, spaced by the rf period.

We briefly describe some of the features of the Stanford superconducting rf linac (SCA) on which many of the early FEL experiments were done. The SCA accelerator has been described by Smith [181], and Table 8.2 in Section 8.3 provides a useful compendium of technical information about the beam as well as the associated FEL. This accelerator has many unique features which make it specially suitable to FEL work.

The accelerating cavities are niobium at 1.9 K, having unloaded $Q > 10^8$ at 1.3 GHz. The rf amplitudes and phases in the accelerating structures are regulated to 0.01% and 0.1° by feedback. The electron gun works at 100 kV and is apertured so that the beam emittance there is 5π mm-mrad; it is accelerated to 2 MV in a capture section. The remaining accelerating sections boost the energy to 20–70 MV. One rather important difference between the SCA and a conventional rf linac is that the rf fields can be maintained much longer in the superconducting cavities—about 2 msec (instead of the typical ≈ 20 μsec) at 2% duty cycle. This is a very desirable feature for an FEL oscillator where the gain is low. Typically the radiation bounce time in the long resonator is ~ 100 nsec, and so 20 μsec of rf provides only 200 passes, which, as we saw in Chapter 3, is barely enough time for the radiation to reach steady state. The peak current in a 3.2-psec current pulse is 1.5 A, the beam energy width is < 20 kV, and the net emittance is $< 0.1\pi$ mm-mrad. The linac efficiency is typically $\approx 50\%$, with good prospects for improvement.

The rf linac is an expensive source, requiring the facilities of an accelerator laboratory. While this provides access to many sophisticated, shared diagnostics and data-handling devices, the facility may be too expensive for FEL research and applications alone. Possibly a less-costly linac devoted exclusively to FEL work can be developed in the future. Meanwhile, there has been attention given to improving two other, less complex high-energy sources, the microtron and the Van de Graaff.

Microtron

A relation of the rf linac is the microtron, a compact electron-beam source in the range of 10–30 MV. An entire book has been written on the microtron accelerator [4]. Recently, improvements in beam current have been made, and two research groups—one at Frascati and the other at Bell Laboratories—have FEL microtron projects underway. The specifications of the Frascati microtron are given in Table 8.1.

The microtron is a very compact accelerator: the magnetic field is not high (1–2 kG is typical), and the chamber is only a meter or two in diameter with a height of about $\frac{1}{2}$ m. The electrons, emitted by a hot cathode positioned near the waveguide, are accelerated by an rf field produced by a klystron or magnetron rf source (≈ 3 GHz). The successive electron orbits, expanding radially as the energy is increased in the nearly uniform vertical magnetic field, are shown in Fig. 8.1. The pulse and micropulse structure of the microtron is the same as for the rf linac. The current is extracted from the microtron via a channel made of soft iron.

Table 8.1 Parameters of the Frascati microtron [33]

Electron-beam energy	20 MV
Average pulse current	350 mA
Peak bunch current	6.5 A
Pulse duration	12 μsec
Energy spread	0.12%
Vertical emittance	3 mm-mrad
Horizontal emittance	6 mm-mrad
Klystron peak power	15 MW
Klystron average power	30 KW
Magnet diameter	106 cm
Bunch length	7 mm
Repetition rate	150 Hz

Electrons are accelerated in a constant-frequency rf field. Synchronism of electron motion with the accelerating field results from the fact that each successive orbit is longer than the preceding one by an integral number of periods of the rf oscillation. Since the revolution period is proportional to the total energy $W = \gamma mc^2$, the increase of energy in a revolution, ΔW, is related to the increase of period, $\Delta\tau$, by

$$\Delta\tau = \frac{2\pi}{ec}\frac{\Delta W}{B_0} \qquad (8.1)$$

We choose $\Delta\tau = \tau_{rf}$, where τ_{rf} is the period of the rf field, for synchronism. ΔW is obtained by passing the electron through the strong rf field in the waveguide. In the microtron, unlike other cyclic accelerators, the energy gain per revolution is large, and the number of revolutions is small—e.g.,

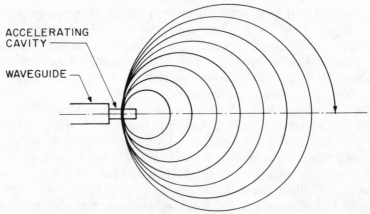

ACCELERATING
CAVITY

WAVEGUIDE

Figure 8.1 Electron orbits in a microtron; the magnetic field is normal to the paper.

10–20. Considerable effort has been devoted to the development of a compact, high-current injector.

The microtron offers a compact and relatively low-cost alternative to the rf linac accelerator, but we sacrifice a certain amount of beam quality.

Van de Graaff

A comprehensive FEL–Van de Graaff project is underway at the University of California, Santa Barbara [67] as part of a two-stage FEL effort (Section 8.5). The first objective is to generate a high-quality electron beam, ≈ 3 MV, 2 A. An electron gun (Hughes Corporation) has been operated which delivers a beam limited only by the emittance from a thermionic source (∼ 1 mm-mrad at 10 kV). Initially, the project aims to demonstrate high energy recovery of the beam, and operation of a submillimeter FEL.

The layout of the facility and the current-balance accounting are shown in Fig. 8.2. The Van de Graaff is powered by a generator coupled to a motor by an insulated shaft, and the electrostatic accelerator is charged at a rate ≈ 500 μA. Extraction of a higher current by the electron beam—viz., 2 A —would draw down the terminal voltage (due to finite capacitance) at ∼ 10 kV/μsec were the electron beam not collected through a decelerating current-recovery terminal. If δ is the fraction of current recovered, then the capacitive decay of the terminal voltage will be $(1 - \delta) \times 10$ kV/μsec. A goal of the project is to pulse the electron gun for 100 μsec, 2 A, at 10-Hz repetition rate.

In addition to matching the gun optics to the electron beam, the beam must also be matched into the collector electron optics. Extensive modeling of the electron trajectories there has been undertaken, using techniques developed by W. B. Hermannsfeldt. Under optimum conditions, Elias [67] has demonstrated that > 99% of the electron current can be recovered, in the absence of the undulator. A good summary of research on electron-gun and depressed-collector techniques suitable for a "low"-voltage FEL has been presented by Dolezal [59].

Storage Ring

The main components of the storage ring include a set of bending magnets aligned in a closed loop, interspersed with a set of quadrupole magnets for focusing, together creating stable orbits. Current is injected into the ring from an external accelerator, and is maintained for a long interval by an rf cavity system which compensates for the loss of energy from synchrotron emission. The rf frequency is an integer multiple of the revolution frequency. For every electron energy, there is a corresponding stable orbit, at least within a range (the energy acceptance) determined by the machine design.

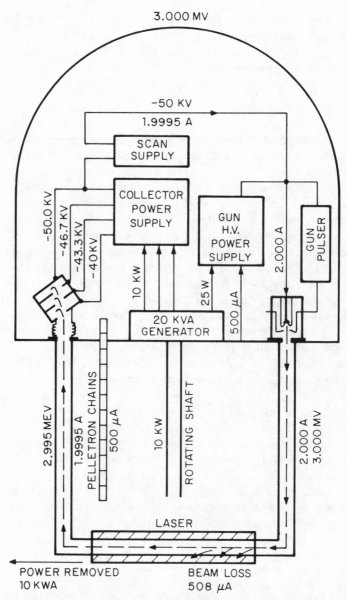

Figure 8.2 Layout of the accelerator and beam recovery system of the UCSB Van de Graaff accelerator [67]. © 1982 Addison-Wesley.

The undulator is inserted in a straight section, usually without important changes occurring in the electron dynamics.

Individual electrons, not in an optimum relationship to the ideal orbit or energy relative to the rf field, will oscillate about some equilibrium position: transverse (betatron) and/or longitudinal (synchrotron) oscillation is possible. The transverse restoring force comes from the quadrupole magnets, while the longitudinal interaction is with the rf field. The motion is damped in either case by synchrotron emission, causing a radiative loss U_s per revolution given by

$$U_s = \frac{4\pi}{3} \frac{r_0}{R} \gamma^4 mc^2 \tag{8.2}$$

where R is the radius of curvature of the electron orbit. This establishes an energy decay time for synchrotron radiation, τ_{syn}. An electron with excessive energy radiates more than the "ideal" (resonant or synchronous) particle, and vice versa; hence an energy deviation is damped eventually. This applies to the transverse oscillations as well, because the emission of a synchrotron photon implies the loss of both transverse and longitudinal momentum.

Storage rings are capable of impressive performance: an optimistic future set of parameters might be $W \approx 1$ GV, $\tau_{syn} \approx 7$ msec, $\varepsilon \approx 0.02$ mm-mrad, $I_\rho > 10\text{A/pulse}$. In the storage ring, the electron-beam emittance does not vary as $1/\gamma$, because of the synchrotron radiation. Scattering of momentum among the electrons may occasionally place a particle outside the momentum acceptance of the ring, resulting in loss: this effect establishes a limit on the beam density, current, and emittance (Touschek effect) that can be sustained for a given time (a typical Touschek time might be ~ 1 hr). A permissible range of electron energies that can be accommodated in the ring might be $\approx \pm 5\%$, but if all the electrons in the beam are to fit as a Gaussian distribution within this acceptance without loss, the width σ_ω of the distribution should be considerably less, say $< 1\%$ [198]. The advantage of the ring is its high current density and low emittance, providing the opportunity for very short-wavelength FEL operation.

Diagnostics

An important part of the accelerator facility is a battery of specialized diagnostics. Electron-beam position, profile, energy, spectrum, and emittance are among the many properties requiring measurement before or during the FEL operation. Most of these involve remote sensing and positioning. The beam location can be monitored at several positions along the beam line by movable fluorescent screens monitored by television. Any pair of screens will help define the trajectory, while the profile is obtained from the optical TV image. Emittance can be determined by fitting the beam profile with a parabola, using the current in the quadrupole magnets as an independent variable. Bending magnets are used for the energy spectrometer, together with a secondary-emission monitor.

The large mirror spacing, together with the complexity of the mechanical structure and the environmental conditions (e.g., vacuum), causes problems regarding remote cavity alignment, cavity length variation, and stability. The spacing must be set to an accuracy of a few microns, and the mirrors should be aligned within 10^{-5} rad for IR conditions. The center of the mirrors must fall on the magnetic axis of the undulator. It is desirable to lock the mirror spacing to the electron bunch separation and compensate for thermal variation. Some FELs have used a control system consisting of a laser interferometer together with a micrometer translation stage and stepping motor. As the electron bunch length is short compared with the round-trip bounce time of the cavity radiation, the resonator Q can be found by observing the decay of stored light in the resonator directly (the Q of a Fabry-Perot cavity is $\sim 2\pi L_c/\lambda_s \alpha_L$, where α_L is the fractional power loss per pass).

8.3   Experiments Using Uniform Undulators

The very first FEL experiments were done at Stanford, using the high-quality SCA linear accelerator [65, 54]. Typical experimental parameters, appropriate to the more recent work, are listed in Table 8.2.

The undulator was a superconducting bifilar helix, while an axial field of 1 kG was used to move the electron beam around the mirrors (Fig. 1.1). The first paper [65] reported the spontaneous emission at 10.6 μm from the undulator when $W \approx 24$ MV, together with 7% amplification of a 140-kW/cm^2 TEA CO$_2$ laser beam, obtaining the now familiar asymmetrical gain curve. That experiment has been repeated using the ACO storage ring operating at 150 MV [55], and in Fig. 8.3 is shown a very accurate comparison of the FEL gain with the derivative of the spontaneous spectrum, obtained using an argon laser (4880 Å). After increasing the SCA pulse current from 70 mA to 2.6 A, it was possible to demonstrate laser oscillation at 3.4 μm using energy 43.5 MV [54]. The peak power represented an enhancement $\sim 10^8$ above the spontaneous emission level. With a mirror transmission of 1.5%, the intercavity power was ≈ 500 kW. Figure 8.4 shows a comparison of the power spectrum below ($\delta\lambda/\lambda \sim 1/N$) and above the oscillation threshold, demonstrating appreciable line narrowing due to the optical resonator. All experimentation was in the two-wave FEL regime, since $\omega_p T \sim 0.1$.

Table 8.2 FEL oscillator parameters at Stanford University [60]

Beam:	
Beam energy	43 MV
Total γ	85
Parallel γ	69
Peak current	1.3 A
Pulse length	1 mm
Pulse separation	25.4 m
Beam radius	0.25 mm
Undulator:	
Period	3.3 cm
Helical-field amplitude	2.3 kG
Length	5.3 m
Resonator:	
Length	12.7 m
Round-trip resonator losses	2.8%
Resonator coupling loss	1.5%
Wavelength	3.3 μm
Spot size	0.167 cm
Rayleigh length	2.7 m
Beam filling factor	0.017
Spontaneous-radiation loss factor	0.05
Mirror radii	7.5 m

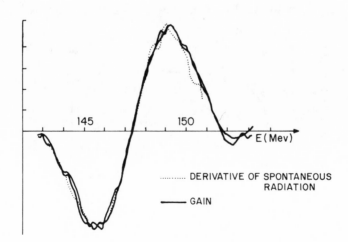

Figure 8.3 Comparison of measured FEL gain (argon laser, 4880 Å) with two separate runs (solid lines) overlaid on the derivative of the spontaneous-emission spectrum (dotted line); the peak gain is 3×10^{-4} [55]. A superconducting undulator having $l = 4$ cm was used. © 1982 North-Holland.

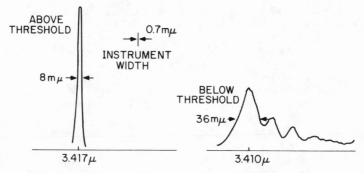

Figure 8.4 Spectral narrowing of the FEL radiation above and below oscillation threshold [54]. © 1977 APS.

Additional experimental results from the Stanford group [60, 25] are particularly valuable as a guide to theoretical reliability. The first point, shown in Fig. 8.5, has to do with the oscillator turn-on time. Not only is this surprisingly long [≈ 30 μsec, several hundred optical passes], but the turn-on time and the power output are quite sensitive to the cavity length (Fig. 8.6). Theoretical understanding of these effects, discussed in Chapter 3, now seems to be in a satisfactory state. A study of the risetime of the radiation level at the beginning of the oscillator pulse permitted an estimate for the small-signal gain per pass, which was ≈ 6–10%, appreciably smaller than would be calculated from Eq. (3.30) on account of filling-factor and finite-pulse-length effects.

Time-resolved measurements of the electron and optical spectra were made. The width of the electron momentum distribution grows to ~ 1% as

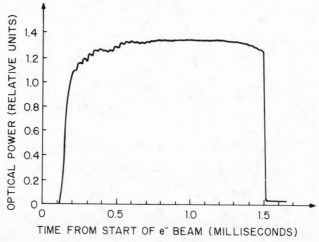

Figure 8.5 FEL radiation pulse shape, showing long startup and risetime [60]. ©1982 Addison-Wesley.

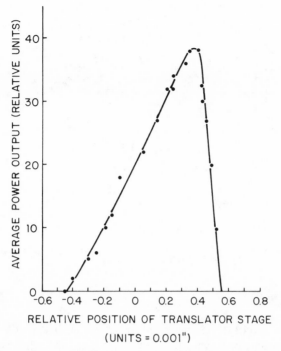

Figure 8.6 Dependence of FEL power upon cavity detuning length [60]. © 1982 Addison-Wesley.

the laser is turned on (an example from an early run is provided in Fig. 8.7, where the centroid is displaced ~ 0.1% from the initial electron energy). The spectrum shows an asymmetry which is suggestive of the theoretical result (Fig. 3.6). The mean optical wavelength remains essentially constant after the initial laser startup, but there are still certain features of the optical pulse that remain to be understood experimentally. In the most recent experiment [25], [206] the optical pulse width was observed using an autocorrelation scheme and second-harmonic generation in a LiNbO$_3$ crystal. When the cavity length was exactly synchronous with the bunch spacing, an FEL pulse width of 1.5 psec was observed, at peak power ≈ 400 kW. When the resonator length was changed, an increase in the optical pulse width and a decrease in the spectral width were measured.

In connection with research on the SCA accelerator at Stanford, mention should be made of the successful operation of the TRW multicomponent undulator FEL oscillator; this experiment was very well diagnosed and well instrumented for control purposes [175]. About 10 W of average power was obtained at 1.6 μm with uniform and nonuniform versions of the undulator. Also, intense third-harmonic radiation was found at ≈ 0.5 μm during the oscillation. As might be expected, the short-wavelength oscillation showed extreme sensitivity to the cavity-length adjustment.

ELECTRON ENERGY SPECTRUM

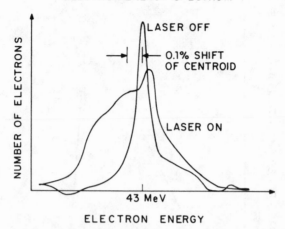

Figure 8.7 Influence of the FEL upon the electron energy spectrum (Madey; reported in [46]. © 1977 Addison-Wesley.

8.4 *Storage-Ring FELs*

FEL research using high-energy electron beams has had two aims from the inception: shorten the wavelength and improve the efficiency. We postpone the latter topic to Section 8.6. The matter of reducing λ_s has technical ramifications which involve more than merely increasing the accelerator energy. One must, of course, hold the gain up [Eq. (3.30) or (4.57)]. Given the low filling factor of a filamentary electron beam in a long Fabry-Perot resonator, it is reasonable to ask for perhaps 100% gain from the formula quoted above. As $a_\omega \sim 1$, the scaling of gain to shorter wavelength is unfavorable, and, apart from improving the undulator design and length, the only recourse is to ask for higher beam current density at low emittance. Madey [128] has commented that the existing SCA facility has operated very near the shortest possible fundamental wavelength for an FEL which uses a linear accelerator (1–3 μm). Concentrating the electron beam, to enhance gain, will cause deterioration of beam emittance and energy spread. On the other hand, the storage ring offers the promise of containing an electron current density orders of magnitude above the linear-accelerator capability, because the limiting current-handling physics of the storage ring —synchrotron radiation and quantum fluctuations—is altogether different from that in the rf linac. Using the high-quality electron beam provided by the storage ring, it has become possible to contemplate FEL operation in the UV, even with conventional undulator designs.

The basic idea in the storage-ring FEL is to recycle the electrons through the undulator over and over again, replenishing the energy radiated in expensive short-wavelength photons with cheap rf energy from the cavity.

The difficulty is that the electrons emerge from the undulator with not only decreased energy on the average, but, more importantly, an increased spread in energies. The Madey gain-spread theorem—refined and generalized over the years [108–110]—shows that electrons emerging from the undulator have a larger energy spread than those which enter, owing to the FEL gain process. A secular growth in this energy spread would occur upon recycling the electrons through the undulator many times, but this is restrained by the synchrotron radiation damping. If the undulator were switched off, electrons in the higher-energy tail of the perturbed electron distribution would radiate more, while those on the lower-energy side would emit less synchrotron light, and eventually the distribution should narrow ["cool"] to its initial value. The characteristic time for reaching a steady state is set by the synchrotron-radiation rate. Renieri [161] first showed that as a consequence, the average power P_L that could be obtained from a uniform undulator FEL in a storage ring should be

$$P_L \approx \frac{1}{2N} U_s \langle I_b \rangle \tag{8.3}$$

The factor $1/2N$ arises from the FEL gain or efficiency, and using the gain-spread theorem, it also must correspond to the energy acceptance of the ring, (viz., the factor $1/2N$ can be replaced roughly by σ_ω). More ingenuity with the undulator configuration, together with additional theoretical effort, has not altered this important conclusion.

Equation (8.3) suggests that the highest power is obtainable from a storage ring with high energy-spread acceptance. However, using a uniform undulator, this quantity would be $\sim 1/2N$, and so it might appear that only a small value for N is compatible with favorable power performance [alas, the gain would then be unacceptably low!]. Fortunately, the incompatibility of power and gain can be broken through the use of a transverse-gradient undulator [128]. It should be possible to design a storage-ring FEL with reasonable power output and adequate gain over a broad range of wavelengths. If, for example, a typical synchrotron power level emitted by the ring is ≈ 20 kW, one can expect the associated FEL to generate about 200 W ($\sigma_\omega \sim 1\%$) in cw operation. The synchrotron energy loss is ≈ 1000–$10,000$ eV per electron per revolution. To increase P_L, an auxiliary undulator can be introduced in the ring to enhance the total spontaneous radiation; in this way, FEL power might be brought to the kilowatt level [153].

It would appear that best FEL results would motivate operation at high energy, as the synchrotron losses are then highest. On the other hand, high-energy electrons passing through the undulator will emit many harmonics as spontaneous radiation in the UV region, and this will cause mirror degradation. Figure 8.8 shows the harmonic radiation calculated for

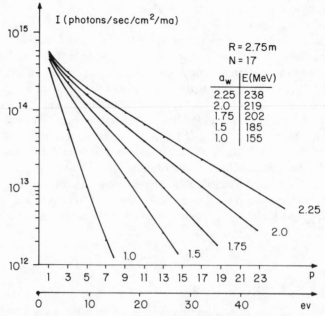

Figure 8.8 Harmonic radiation levels calculated for electrons in the linearly polarized undulator NOEL on the ACO ring [28].

the NOEL dipole undulator on the ACO storage ring at Orsay. This study encourages operation at minimum ring energy, at least until a better mirror surface is developed.

The first storage-ring FEL oscillator in the visible has been reported [29] by the Stanford-Orsay research group, using the ACO ring (Table 8.3). The ring energy was reduced for this experiment to 160 MV, with current 100 mA, in order to limit mirror degradation from the strong UV synchrotron-radiation flux. Ordinarily, the energy would have been higher, viz. ≈ 240 MV. An optical-klystron undulator was used to enhance the undulator gain, permitting oscillation ($G \sim 10^{-3}$); the undulator assembly consisted of two linearly polarized sections having seven periods each, $l_0 = 7$ cm, separated by a dispersive section. Figure 8.9 shows the spontaneous-emission spectrum measured from the optical-klystron undulator: the numerous "oscillations" are caused by the dispersive section, and become more closely spaced as the dispersion is increased (the periodicity in units of $\Delta\lambda/\lambda$ is $\sim 1/N_d$). The undulator was operated with a_ω up to 1.2. The mirrors were 24-layer dielectric [TiO_2, SiO_2], with reflectance of 99.97% and transmission of 3×10^{-5}. Oscillation was sustained in the TEM_{00} mode for about one hour at a time, at three wavelengths near 6500 Å, with peak power level ≈ 60 mW and intercavity peak power $\approx 2kW$. The average power (~ 100 μW) may be compared with the synchrotron power of ≈ 3.1 W; the group estimates

Table 8.3 ACO storage-ring FEL [29]

Energy	160–166 MV
Circumference	22 m
Beam current	16–100 mA
Bunch length	$\frac{1}{2}$–1 nsec
Transverse size	0.3–0.5 mm
Cavity L_c	5.5 m
Mirror radius	3.0 m
Cavity losses	7×10^{-4}
Rayleigh range	1 m
Energy spread	$\sim 10^{-3}$
Length of Optical Klystron	133 cm

that the output power is about $\frac{1}{2}$ the Renieri limit for the ring. The time-averaged spectral linewidth was ≈ 2 Å, but observations during a short interval gave a line width < 0.3 Å.

In another interesting experiment [164], the interaction of an external laser with the storage-ring FEL system was found to change the electron-bunch length. At low current, the laser caused the bunch to lengthen, as might be expected, since the laser interaction with the undulator results in an increase in the electron energy spread. However, at high current the bunching was observed to improve. What may happen is that the FEL interaction tends to stabilize a minor instability in the ring which on its own manages to spread out the bunches.

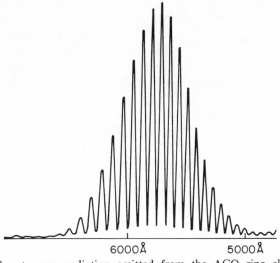

Figure 8.9 Spontaneous radiation emitted from the ACO ring electron beam, passing through the optical-klystron undulator [28].

Progress in reducing the wavelength depends on improvement in mirror coatings and control of the UV ambient radiation. Use of a helical undulator would contribute fewer harmonics, and improved Brewster windows would shield the mirrors from the UV flux. The ACO group found that conventional Brewster windows darkened very rapidly, and therefore a special type of material is called for.

Operation of a storage-ring FEL producing several watts in the range 500–200 Å is contemplated by the construction of a high-current, 1-GV facility at Stanford. With space for a high-gain 15–20-m uniform undulator, the requirement on mirror reflectance becomes liberal. Interestingly enough, there has been considerable effort to develop high-reflectivity materials for use in the soft X-ray region [21]. Layered microstructures, consisting of many alternating layers of high-Z and low-Z materials having thickness of several angstroms, can give $> 50\%$ reflectance at normal incidence for λ in the 200-Å range. Assessment of the feasibility of an UV FEL, done at the Brookhaven meeting of September 1983, found that there is no fundamental obstacle to building a storage ring which will give adequate gain at wavelength ≈ 500 Å. What is noteworthy is that, taking $a_\omega \approx 2$, the device may fall within the strong-pump regime [see Section 4.3]. Assuming realistic emittance and energy-spread criteria, the net power gain per pass could be > 10. To be sure, more careful attention must be devoted to optimizing the vertical and horizontal beam radii with respect to the optical mode than in the IR, and the emittance must be very low (10^{-2} mm-mrad) at high pulse current (100–200 A).

A short-wavelength, high-gain FEL operating at $\lambda \approx 30$ Å has been proposed by Pellegrini. Mirrors could be dispensed with by operating in the strong-pump regime of growing exponential instability; indeed, this X-ray FEL would be a superfluorescent amplifier several e-folding lengths in extent, driven to saturation. A feature of the design is that a storage ring would provide a high-quality beam which is "occasionally" diverted through a long [$N \approx 500$] undulator, after which it is reinjected into the ring for cooling. This feature permits a small undulator gap and period [$l_0 \approx 0.5$ cm] to be used.

At very short wavelengths, the effect of the discrete photons upon FEL theory is still an active research area. Some insight may be gained by estimating the number of electrons and photons in a characteristic volume, which we call V_0. The radiation emerges in a beam of roughly the same dimension as the electron beam, having area $\pi w_0^2 = \lambda Z_R \sim \lambda L$, where Z_R is the Rayleigh range (approximately the undulator length). The length of the characteristic volume is $\sim N\lambda$; this "slippage" distance is enough for the electron to sense the FEL bandwidth and coherence, including the important range of induced emission and absorption effects. Therefore, $V_0 \sim N\lambda^2 L$, which shows why questions about statistics must arise as λ decreases. However, for a storage-ring UV FEL at $\lambda = 500$ Å, $V_0 = 10^{-5}$

cm^3, and there are $\langle n \rangle \sim 10^7$ electrons in V_0. The transition probability near saturated power level in a storage-ring FEL is roughly 1000 times the spontaneous rate, and one finds about one photon is emitted per electron (a situation similar to the atomic laser) during laser operation. The photon number in V_0 is large both during the FEL start-up and in the steady state. Shot-noise effects arise from δn and represent an important contribution to the FEL startup.

8.5   *The Two-Stage FEL*

The two-stage FEL, proposed by Elias [66], is an approach to reducing the beam energy needed to generate short wavelengths. The first stage uses an electron beam having energy of a few megavolts together with an undulator having a period of a few centimeters to produce a submillimeter, coherent FEL oscillation. This EM radiation is to be contained in a high-Q resonator, where its intensity should reach very high level—e.g., > 100 MW/cm^2. The submillimeter radiation becomes a pump on the same electron beam (Fig. 8.10), and the second-stage radiation is regenerated in another resonator. As the first stimulated scattering interaction gives us $\lambda_{s1} = l_0/2\gamma_{\parallel}^2$, and the second scattering interaction gives $\lambda_{s2} = \lambda_{s1}/4\gamma_{\parallel}^2$, the wavelength of the second-stage FEL is roughly $l_0/8\gamma_{\parallel}^4$. Useful amounts of radiation at λ_{s2} can be produced only when the EM field of λ_{s1} is very intense. The electron beam enters the second-stage region first, where the highest-quality electron beam is required. As the beam passes through the undulator, the strong optical field causes substantial momentum spread in that location.

The first interaction would take place in a quasi-optical Fabry-Perot cavity, where an annular electron beam could excite a low-loss TE$_{01}$ mode —also with an annular intensity pattern—in a cylindrical waveguide (Fig. 8.10). Holes in the end mirrors could permit λ_{s2} to leave the submillimeter cavity and yet these would present negligible losses for the TE$_{01}$ mode at λ_{s1}. This design has been the focus of a study by the FEL group at KMS

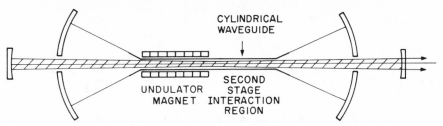

Figure 8.10   Experimental layout of the two-stage FEL [166]. Optical mirrors are at the extreme ends.

Fusion [166]. They have taken the electron-beam energy to be 3 MV, the current in the range 2–20 A, and $(\delta\gamma/\gamma)_{||} < 10^{-4}$; the second-stage region would be about 3 m in length. If the pump wavelength ≈ 1 mm, the second-stage wavelength becomes ≈ 5.3 μm.

In Fig. 8.11 is shown the results of a KMS calculation of the expected power output of the FEL at λ_{s2}. A key quantity is the loss per pass in the second-stage optical system (this includes both resonator and coupling losses), as well as the beam current and the millimeter-pump power. The latter depends on the power supplied from the electron-beam system (Fig. 8.2). The calculation predicts the maximum power that could be obtained, based on an assumption of the expected cavity losses.

As the intercavity power at λ_{s1} will be $\sim 10^8$ W, one might expect that the first-stage operation would cause a large beam energy spread which would make efficient beam-energy recovery difficult. The expected amount

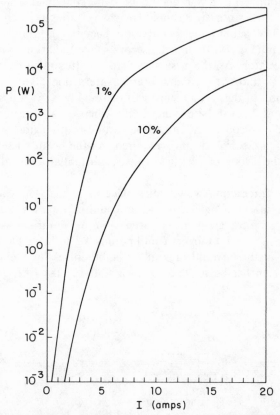

Figure 8.11 Power calculated from a two-stage FEL operated at $W = 3$ MV, $L = 3$ m, $\lambda_{s1} = 1$ mm, $\lambda_{s2} = 5.3$ μm, with $p_{\text{pump}} = 10^7$ I_b, taking the gain as 1% and 10% [166].

of energy spread is of the order of the bucket-height formula [Eq. (4.80)]. Takeda and Segall [188] have suggested that the phase-space displacement method may prove helpful here. Unlike the situation described by Kroll et al. [108], the initial energy spread of the electron distribution would be negligible compared with the bucket height. For appropriate choice of the resonant phase, viz. a reverse-taper undulator, it is predicted that the operation of the first FEL stage can occur at high optical power, but with diminished energy spread and adequate gain, provided the electrons execute a number of synchrotron oscillations in the undulator.

One complication is that the device will operate in a pulsed electron current mode, because the Van de Graaff accelerator system cannot supply the full beam current continuously at constant voltage. This implies—since both the pump EM wave and the short wavelength radiation must build up from spontaneous radiation—that if the accelerator pulse were too short, the configuration might fail to achieve threshold, or might not approach its asymptotic performance. Hence the importance of an excellent beam-current recovery system, which will sustain high current over many microseconds with negligible accelerator voltage droop.

Operation of the submillimeter part of this FEL was reported in 1984 at Santa Barbara. The linearly polarized undulator has $N = 160$, $L = 5.76$ m, $l_0 = 3.6$ cm, $B_\perp = 600$ G, and generates several kW at 380 μm with a 3-MV, 2-A beam. If in fact the submillimeter radiation can be increased to 300 kW, then about 100 W of radiation at 6000 Å is expected from the second stage.

Another approach to the same problem—but one lacking an experimental sponsor—is to use a Raman FEL as a first stage. Here one chooses a different technology, using perhaps a beam of 2 MV at 3 kA, which should generate as much as 500 MW at $\frac{1}{2}$ mm in the first stage. Injection of this pump power into a second stage consisting of a 5-MV, 1-kA beam might produce a comparable power level near 1 μm, assuming that one "tapers" the amplitude of the $\frac{1}{2}$-mm pump wave by channeling it in a waveguide having slowly changing wall spacing. In contrast, to produce the same wavelength, a 100-MV single-stage facility would be required. By now the reader will appreciate that it is the beam quality—particularly that of the second stage—which will make or break this proposal; the diagnostics and handling of intense, low-γ beams is still in its infancy.

Can we reach very short wavelength, perhaps 100 Å or less, using a two-stage free-electron laser? Rather than a high-energy storage ring using a conventional undulator, we might consider a modest-energy, but very high-quality electron beam [e.g., $(\delta\gamma/\gamma)_\parallel \sim 10^{-5}$], pumped by a short-wavelength, high-power optical laser. Take the example of a 1.06-μm Nd-glass laser, which is capable of pulse power ~ 1 TW. With careful attention to the laser oscillator, one might contemplate a linewidth as low as 0.1 cm^{-1}, i.e., $\delta\lambda/\lambda \sim 10^{-5}$, whereupon the EM wave from this optical laser could be used as an undulator of perhaps 10^5 periods and yet retain fair coherence. The magnetic pump field of the 1-μm radiation is of the order of several megagauss, yet the quiver velocity is low and only ≈ 2 MV is required for the beam energy. It is not unreasonable to expect that the parallel energy spread requirement could be met by a small beam having a current ≈ 2 A [the undulator inhomogeneous broadening is not given by Eq. 5.29, as the optical-beam intensity can be nearly uniform over the electron beam]. Substitution into the classical gain formula indicates that rather high power gain is feasible, and that one could imagine this as a potential superfluorescent amplifier for synchrotron radiation: no mirrors required! A more detailed study shows the scheme to be less attractive [138, 58]. Most importantly, the Compton recoil effect appreciably reduces the gain. We can understand this by returning to Eq. (2.45): the wavelength shift of the radiation due to the electron recoil, $\Delta\lambda_c$, is such that $\Delta\lambda_c/\lambda \sim 2\hbar\omega/\gamma mc^2 = 2\lambda_c/\gamma\lambda$, where λ_c is the Compton wavelength h/mc. The gain will deteriorate if $\Delta\lambda_c/\lambda$ is larger than the FEL gain bandwidth, $1/N$. Hence, the inequality for preserving FEL gain against Compton recoil is $1 > \lambda_c N/\gamma\lambda$. The two-stage FEL with the "optical" undulator does not satisfy this inequality, and hence its gain is smaller than the classical formula indicates; the storage-ring FEL, with its large γ and smaller N, satisfies the Compton inequality. On the other hand, Compton recoil allows some liberalization of the beam momentum-spread requirement.

8.6 *Efficiency Enhancement with Nonuniform Undulators*

Certainly one of the most challenging tasks—not only for FEL theory, but with regard to the viability of the FEL concept—has been the demonstration of the variable-parameter undulator FEL. Three groups [TRW, Boeing–Math Sciences (B/MS), and Los Alamos (LANL)] have participated in this project; research summaries are to be found in papers by Edighoffer et al. [62], Slater et al. [177], and Warren [197], respectively. The progress has been excellent, and the principal expectations are borne out by experimental data. We shall summarize this work below, concentrating on the B/MS and LANL results, obtained with similar equipment.

The strategy is to use a high-power 10.6-μm CO_2 laser to trap electrons in the ponderomotive buckets, demonstrating that energy can be extracted from the relativistic electrons and transferred to the optical wave, as gain, using a nonuniform undulator. The allowable emittance of the electron beam can be set by two rather different requirements. First, one can ask that the "bucket height" [Eq. (4.80)] be larger than the beam energy spread due to emittance [Eq. (5.31)]; this condition will permit the electrons to be trapped into the ponderomotive buckets as if they were cold. As is customary, $a_w \sim 1$ and the undulator contribution to inhomogeneous broadening is not appreciable, given the small beam radius. Consulting Eq. (3.34), saturation will occur when the amplitude of the optical field is

$$e_s \sim \frac{\gamma^2}{a_w L}\left(\frac{\Delta\gamma_r}{\gamma}\right) \tag{8.4}$$

where $\Delta\gamma_r$ is the change of resonant energy along the undulator. It turns out [158] that the energy-spread requirements at small-signal and saturated conditions are roughly the same for the linear-taper undulator. The second criterion is set by rather different conditions: to obtain maximum gain, it is reasonable to ask for spatial overlap of the optical- and electron-beam envelopes, given $r_b \sim w_0$. The optical beam expands by diffraction and the electron beam expands by divergence $(v_\perp/v)$, and by matching these one finds the condition $\epsilon_N < \gamma\lambda_s/2$ [158]. This is more restrictive at short wavelength.

Table 8.4 is a set of hardware parameters appropriate to the LANL project. Of particular note is the CO_2 laser intensity, as well as the undulator, made up of linear permanent dipoles. Under design conditions

Table 8.4 Nominal values of LANL parameters [197]

Optical:	
Wavelength	10.6 μm
Power	50–900 MW
Rayleigh length	40 cm
Focal-spot radius	0.16 cm
Pulse length	5–8 nsec
Electron beam [*linac*]:	
Average current	0.3 A
Peak current	7 A
Electron energy	19–22 MV
Total energy spread	$\frac{1}{2}\%$
Emittance	2π mm-mrad
Focal spot, undulator center	0.05 cm
Micropulse length	35 psec
Micropulse spacing	770 psec
Undulator:	
Length	100 cm
Period (max/min)	2.7/2.4 cm
Amplitude	3 kG
Taper	12%
Length of entrance/exit zone	5 cm
Performance (*design*):	
Max energy extraction, 900 MW	3.7%
Max optical gain, 500 MW	2.4%
Max optical gain, small signal	3.1%

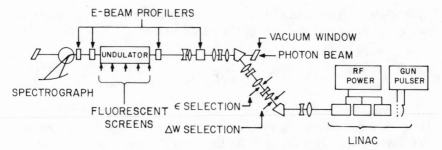

Figure 8.12 Accelerator and laser facility at B/MS.

[37], about half of the electrons should become trapped and decelerated by 7% in energy, providing an average energy extraction in the range 3–4%. Both the B/MS and LANL groups have taken extraordinary measures to align the electron beam and inject it into the undulator under conditions of low beam emittance and energy spread (Fig. 8.12). Even the assembly of the undulator requires a meticulous matching of the individual samarium-cobalt elements to assure adequate uniformity. The long bunch length means that lethargy effects are not important in the interpretation of the experimental results.

We can summarize the major experimental conclusions by inspecting Figs. 8.13, 8.14, and 8.15. The first (B/MS) shows the electron-beam energy spectrum before (solid line) and during the time of peak CO_2 laser intensity (dotted line, 500 MW). The shift of the beam-energy centroid is 4% in this example [91]. One-dimensional numerical calculation has been able to predict the electron energy spectrum at the undulator output (dashed line) assuming that the electron and optical beam are optimally aligned and

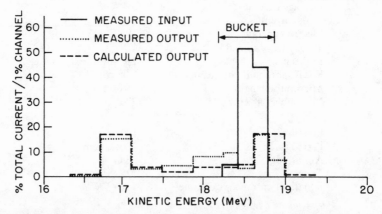

Figure 8.13 Electron spectrum at the entry (solid line) and outlet (dotted line) of a nonuniform undulator [91]. The dashed line indicates a computation. Optical power, $\frac{1}{2}$ GW; $I_b = 160$ mA, peak.

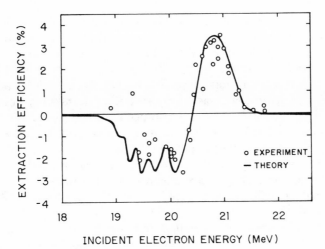

Figure 8.14 Extraction of electron-beam energy in a nonuniform-undulator experiment at LANL [197]. © 1983 IEEE.

focused, and that the optical beam is diffraction-limited. Further computational refinements may require modeling in the transverse direction.

In Fig. 8.14 (LANL), the energy extracted from the electrons depends on the initial beam energy as one might expect qualitatively; the details are due to the complications of the tapered undulator and the appreciable energy extraction. Maximum extraction corresponds to electrons entering the undu-

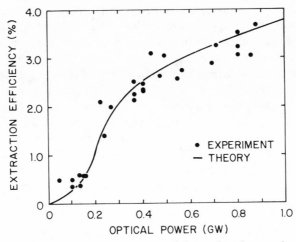

Figure 8.15 Dependence of energy extracted from the electron beam (using a nonuniform undulator) upon laser power [197]. © 1983 IEEE.

lator at the resonant energy for the buckets at that position. An increase of electron energy (negative extraction efficiency) results from acceleration at the expense of the laser field. Finally, in Fig. 8.15 one sees how the energy extracted from the electron beam depends on the optical power: the gain depends on the slope of this curve. The buckets are well developed above the knee of the curve, but—as energy spectra confirm—they disappear at lower optical intensity. This result also demonstrates the sensitivity of the nonuniform undulator to the input optical power level. For a beam emittance of 2π mm-mrad, the momentum distribution was entirely contained within the ponderomotive well formed by the undulator and the laser wave. When the beam energy spread was increased, a blurring-out of the energy spectra resulted, although no change in the extraction efficiency resulted up to $(\delta\gamma/\gamma)_{\parallel} \approx 2\%$.

Direct measurement of the optical gain is more difficult because of the short micropulses. The TRW group has measured a gain $\approx 1.5\%$ at 10.6 μm in a uniform-undulator experiment. In that case, the laser power was lower by a factor of 10 and the electron-beam quality was worse than above, so that trapping was not as satisfactory. More recently, Newnam et al. [144] have reported optical gain of one percent at the LANL facility, using a tapered undulator.

Whereas the amplifier experiments were carried out with a pulsed current of a few amperes, recent modifications have increased the beam current to ≈ 100 A with no deterioration in beam quality. This has been possible using an rf feedback control system attached to the linac bunchers. The purpose of this accelerator upgrade is to create a high-power, high-energy extraction FEL oscillator, to be operated at 0.5 μm by B/MS and at 10.6 μm by LANL. The addition of optical feedback will permit investigation of the sideband instability, a particularly interesting situation in that there are only about 1.5 synchrotron orbits in the LANL undulator. An intercavity filter and several nonuniform undulator variations will provide data pertinent to the sideband question.

A recent oscillator experiment reported by Edighoffer et al. [63] has used a multicomponent undulator, combining the features of the optical klystron (large small-signal gain) with those of the tapered undulator (high extraction efficiency). The undulator consists of eight sections containing fifteen periods, with constant period = 3.6 cm. While the optical-klystron section provides high gain—about 7%—at low signal level, as the optical field increases the magnetic-dispersion section bunches the electrons at the entrance of the tapered section (where only $B_{\perp}$ varies) to provide improved trapping within the tapered region. Using the Stanford SCA accelerator, output at 1.57, 0.8, and 0.53 μm was obtained, yielding pulsed power of 1.2 MW (460 MW or 11 GW/cm^2 inside the resonator) on the fundamental. Based on electron-energy measurements, the efficiency was 1.1% for a

1%-taper undulator and $\approx 0.3\%$ for a uniform undulator; the latter is close to the theoretical expectation.

For high-power oscillators optical damage to the mirrors may become a problem even for initial experiments. Water-cooled dielectric mirrors may be necessary, with the attendant problem of maintaining optical tolerances under thermal stress. To avoid damage to the mirror surface, the cavity length should be expanded so that the FEL mode pattern is spread over a large surface on the mirror. At distance $L_c/2$ from the optical-beam waist $(2w_0)$, the optical-beam radius becomes $\sim L_c w_0/2L_R$, where L_R is the Rayleigh range, $\pi w_0^2/\lambda_s$. If the FEL resonator contains power P_c, and I_D is the radiation-damage intensity limit on the mirror surface, then the resonator length must be at least

$$L_c \geq \left(\frac{8P_c \Delta t L_R}{\lambda_s I_D} \right)^{1/2} \qquad (8.5)$$

[93]. A typical high-power situation might involve a circulating power of 5 GW in 30-psec micropulses, with 10^4 micropulses per macropulse. Taking a Rayleigh range of 1 m, $\lambda_s = \frac{1}{2}$ μm, and a macropulse damage limit of 10 J/cm^2, the required cavity length is ≈ 500 m (this could be reduced with intracavity telescopes). Under these conditions, the mirror curvature radii would be $\sim L_c/2$ and the cavity is nearly concentric. Alignment tolerances are very strict and become more demanding in the direction of shorter wavelength, shorter Rayleigh range, and higher power.

While it might appear that large P_c will result from an efficiency-enhancement scheme, it is much easier to obtain large cavity power by simply increasing the resonator Q, that is, using high mirror reflectivity. Even a simple undulator will then extract enough power from the electron beam to trap the electrons in the ponderomotive buckets. One can avoid difficult problems with optical damage by coupling out a larger fraction of the cavity power.

A useful parameter involved in the design of the resonator is the Fresnel number, $d^2/\lambda_s L$. Both the undulator (gap d, length L) and the mirror system (radius R, length L_c) can be characterized by their Fresnel numbers. The Fresnel number emerges from a diffraction treatment of the waves in the resonator, ordinarily via the Huygens integral; it is the number of Fresnel zones on the aperture. Once the Fresnel number is > 1, losses due to diffraction will drop rapidly (see [14, p. 341] for examples).

Real FELs have a finite-size electron beam—there is axial as well as radial structure. The gain function is therefore three-dimensional, and there is interaction with the three-dimensional mode structure of the Fabry-Perot resonator [69]. The nonplanar waves alter such important results as the Madey theorem, and change the distribution of optical power in the spectrum (for examples, see [50, 52, 130]). There are additional effects peculiar to high-power operation, such as optical self-focusing in the electron beam [108, 191]. Diffraction tends to smooth over many of these phenomena. Numerical study depends to a high degree on specific models, but is nonetheless helpful in understanding important details of FEL operation.

8.7 The Microtron FEL

Two microtron-FEL projects are underway, one at Frascati [33] and the other at Bell Laboratories [167]. The Frascati project is directed toward wavelengths in the range 30–16 μm, whereas the Bell facility is interested in much longer wavelengths, about 100–400 μm. Both groups use similar microtrons and klystron power sources.

The Frascati project has developed a samarium-cobalt element undulator having $l_0 = 5$ cm, $L = 2.25$ m, and $B_\perp \sim 3$–6 kG [$a_w \sim 1$–2]. The resonator length is 6.15 m, and using a confocal configuration they obtain a fundamental mode with diameter $2w_0 = 6.8$ mm. The undulator Fresnel number is $d^2/\lambda_s L \approx 2$, where d is the undulator gap; the diffraction loss per pass in the undulator is then only about 0.1%. The layout of their oscillator experiment is shown in Fig. 8.16.

The group at Bell has chosen to operate at long wavelength in order to use the short coherent pulses of FEL radiation for a variety of problems in solid-state physics (Chapter 1). The microtron (Fig. 8.17)—fabricated by Nucleotronics/Scanditronix—will operate at energy 10–20 MV, providing up to 150 mA of current in 5-A micropulses. The beam area should be about 1 cm. Their large-bore helical undulator, of novel design [170], is quite long (10 m) with 20-cm period and $B_\perp = 400$ G. To reduce power consumption, they have used helically cut iron poles with straight current-carrying conductors (the heat-load reduction with this design is from 800 kW down to only 20 kW). The undulator is located on the axis of a far-IR confocal cavity which has copper mirrors in vacuum. Radiation is to be extracted from the cavity by a hole in one of the mirrors. If all goes well, the gain per pass will be $\approx 40\%$ and a peak power ~ 100 kW should be obtained.

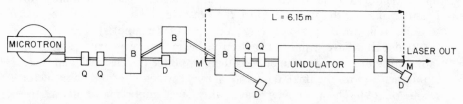

Figure 8.16 Layout of the Frascati microtron-FEL experiment [33]. B is a bending magnet, and Q is a focusing magnet.

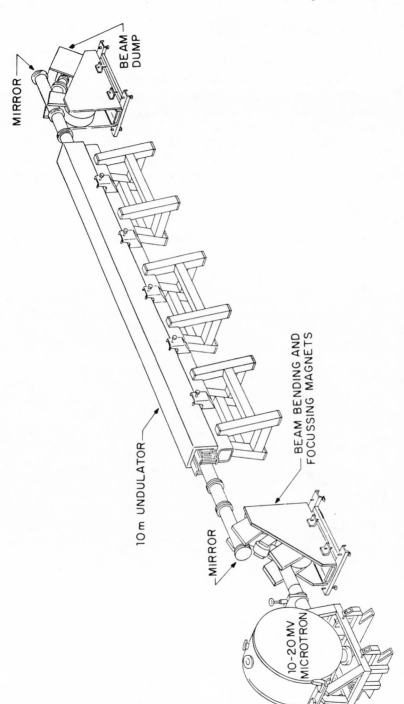

Figure 8.17 The FEL at Bell Laboratories, showing the microtron (left), the two bending and focusing magnets, the undulator (13 m), and two mirrors.

The long wavelength of the Bell FEL and the electron bunch length ≈ 6 mm combine to present a troublesome lethargy problem. The slippage parameter $s = N\lambda_s/l_p$ (defined in Chapter 3) is ≈ 2: therefore the EM wave in the resonator traverses the electron-beam pulse in a fraction of the undulator length, limiting the gain. However, Shaw [168] has pointed out that a narrowband metal mesh filter located in the resonator can resynchronize the optical pulse with the electron bunch. A mesh filter $Q \sim 100$, well within technological capability, is required. There is accordingly reason to expect favorable performance from this type of FEL, down to longer wavelengths where the Raman process begins to enhance the two-wave gain.

Afterword

The growth of understanding and implementation of the free-electron laser has indeed been impressive. One may expect that FELs will soon be established in research and engineering applications serving a wide spectral range. Applicability has been demonstrated from the microwave domain to the soft X-ray region, using a variety of electron (and perhaps positron) accelerators. FELs can compete with atomic lasers even in the visible and near IR when specialized applications requiring high efficiency, high power, or tunability are paramount. At very short wavelength, or at long wavelength (excepting microwaves), there is little competition.

In our survey of FEL theory, we have emphasized simple models. Limiting cases include the "single particle" FELs, namely the two-wave Compton (eq. 3.30) and warm-beam Compton (eq. 4.60), where the gain or growth parameter scales as ω_p^2 or n; on the other hand, there are FELs where the pump field from the undulator is strong or the plasma frequency is sufficiently high, and these exhibit "collective" behavior in which the growth scales nonlinearly with n (eq 4.53). Actual design may require the more accurate travelling wave model which can incorporate the electron orbits (section 6.2), or in situations where the undulator is non-uniform, numerical integration of the generalized pendulum equation (section 4.5). Finally, diffraction or other two-dimensional effects require a more sophisticated numerical study: these cases are outside the scope of this book, but the complexities do not radically modify the concept of FEL operation that has been established.

Experimentally, The FEL and its various refinements have been studied in sufficient detail to validate the concept and to show where possible limitations may apply. At very short wavelength, for example, the upper limit of charge density in the storage ring bunch is crucial for successful FEL operation, but it remains an open question for accelerator physics to resolve. For high power operation, it is still too early to tell whether the sideband instability represents a well-defined limitation or rather only a nuisance. In matters pertaining to mirrors or optical techniques, the FEL will continue to benefit from advances in laser technology, such as the progression of conventional lasers into the vacuum ultraviolet. These and other questions can now be studied at a variety of experimental facilities devoted to FEL research.

This new technology requires some adjustment in our conception of what a laser facility might be. The prevailing viewpoint—that a laser is a device costing several thousand dollars and occupying a corner of a research laboratory—is not appropriate to the FEL. (Of course, some laser facilities, e.g. NOVA, do occupy the large buildings.) Our idea of a coherent source also changed after the invention of the laser: it was no longer an electron tube priced at several hundred dollars and fitting into a small box. The FEL is best viewed as a specialized source for which a portion of an accelerator facility (up to a few million dollars) is dedicated. The applications must be unique and justify the additional demands on space, manpower, and resources. However, cost can be reduced by time sharing. There is some indication that attitudes toward FEL facilities may be changing in this direction.

In the meantime, experimentation proceeds in its traditional role, to establish the validity of theoretical models and to test the limitations of technology. As this is written, an FEL project can secure support for those reasons alone. However, the time is approaching when the applications will determine the pattern of investment. Informed investment requires understanding throughout the wider technical audience of what has been until now an esoteric branch of laser physics. The purpose of this book is to bring FEL physics to the attention of those who will make these decisions.

Bibliography

"Phys. Quan. Elect." is an abbreviation for the series published by Addison-Wesley, edited by Jacobs et al., entitled "The Physics of Quantum Electronics."

"SPIE" stands for a publication by the International Society for Optical Engineering, Bellingham, Washington; it includes the papers presented at the Free Electron Laser Workshop held at Rosario, Washington in June 1983; editors: C.A. Brau, S.F. Jacobs, M.O. Scully.

Conference papers referenced as "Lasers, '82," etc., are published by the Society of Optical and Quantum Electronics, STS Press, McLean, Virginia.

Books and Review Papers

1. R.C. Davidson, "Theory of Nonneutral Plasmas" (Benjamin, New York, 1974).
2. V.L. Granatstein, R.K. Parker, and P. Sprangle, in "CRC Handbook of Laser Science and Technology," M.J. Weber ed., Ch. 4.12, p. 441 (CRC Press, Boca Raton, Fla., 1982).
3. J.D. Jackson, "Classical Electrodynamics," (Wiley, Second Edition 1975), Ch. 14.
4. S.P. Kapitza and V.N. Melekhin, "The Microtron," E.M. Rowe, ed. (Harwood, London, 1978).
5. N.M. Kroll, Phys. Quan. Elect., vol. 5, p. 115, 1978.
6. N.M. Kroll, P. Morton, and M.N. Rosenbluth, J. Quan. Elect., vol. QE-17, p. 1436; also, Phys. Quan. Elect., vol. 7, pp. 89, 113, 147; 1980.
7. J.D. Lawson, "The Physics of Charged-Particle Beams," Oxford, 1977.

8. T.C. Marshall, S.P. Schlesinger, and D.B. McDermott, Adv. Electr. and Electron. Phys., vol. 53, p. 47, 1980.

9. R.B. Miller, "Introduction to the Physics of Intense Charged Particle Beams," Plenum, New York, 1982.

10. C.K.N. Patel, R. Bersohn, C.A. Brau, L. Elias, G. Haller, A. Mooradian, C. Pellegrini, and P.L. Richards, "The Free Electron Laser," report of the subcommittee of the Solid State Sciences Committee, The National Academy, 1982.

11. D. Prosnitz, in "CRC Handbook of Lasers," Ch. 4.11, p. 425 (CRC Press, Boca Raton, Fla., 1982).

12. A. Renieri, in "Developments in High-Power Lasers and Their Applications," Enrico Fermi School 1978, C. Pellegrini, ed., p. 414 (North-Holland, 1981).

13. M. Sands, in "XLVI Corso, Enrico Fermi School," p. 257 (Academic Press, New York, 1971).

14. A.E. Siegman, "An Introduction to Lasers and Masers" (McGraw-Hill, 1971).

15. P. Sprangle, R.A. Smith, and V.L. Granatstein, Infrared and MM Waves, vol. 1, p. 274, 1979.

Publications

16. H. AlAbawi, F.A. Hopf, G.T. Moore, and M.O. Scully, Opt. Comm., vol. 30, p. 235, 1979.

17. M.J. Alguard, R.L. Swent, R.H. Pantell, B.L. Berman, S.D. Bloom, and S. Datz, Phys. Rev. Lett., vol. 42, p. 1148, 1979.

18. P. Avivi, F. Dothan, A. Fruchtman, A. Ljudmirsky, and J.L. Hirshfield, Intl. J. IR and MM Waves, vol. 2, p. 1071, 1981.

19. A. Bambini, A. Renieri, and S. Stenholm, Phys. Rev. A, vol. 19, p. 2013, 1979.

20. G. Barak and N. Rostoker, Phys. Fluids, vol. 26, p. 856, 1983.

21. T.W. Barbee, Jr., in "Low Energy X-Ray Diagnostics," D.T. Attwood and B.L. Henke, eds., vol. 75, p. 131, Amer. Inst. Phys., 1981.

22. G. Bekefi, Appl. Phys. Lett., vol. 40, p. 578, 1982.

23. G. Bekefi, R.E. Shefer, and B.D. Nevins, in "International Conference on Lasers '82," p. 136 (Soc. Optical and Quantum Electronics, 1982).

24. G. Bekefi, R.E. Shefer, and W.W. Destler, Appl. Phys. Lett., vol. 44, 280 (1984).

25. S. Benson, D.A.G. Deacon, J.N. Eckstein, J.M.J. Madey, K. Robinson, T.I. Smith, and R. Taber, J. de Physique, vol. 44, suppl. 2, p. C1-353, 1982.

26. I. Bernstein and J.L. Hirshfield, Phys. Rev. A, vol. 20, 1661, 1979.

27. I.B. Bernstein and L. Friedland, Phys. Rev. A, vol. 23, p. 816, 1981.

28. M. Billardon, D.A.G. Deacon, P. Elleaume, J.M. Ortega, K.E. Robinson, C. Bazin, M. Bergher, J.M.J. Madey, Y. Petroff, and M. Velghe, J. de Physique, vol. 44, suppl. 2, p. C1-29, 1983.

29. M. Billardon, P. Elleaume, J.M. Ortega, C. Bazin, M. Bergher, M. Velghe, Y. Petroff, D.A.G. Deacon, K.E. Robinson, and J.M.J. Madey, Phys. Rev. Lett., vol. 51, p. 1652, 1983.

30. D.S. Birkett, T.C. Marshall, S.P. Schlesinger, and D.B. McDermott, J. Quan. Electr., vol. QE-17, p. 1348, 1981.

31. D.S. Birkett and T.C. Marshall, Phys. Fluids, vol. 24, p. 178, 1981.

32. D.S. Birkett, "A Submillimeter FEL: Experiment, Analysis, and Cavity Mode Theory," Thesis, EE Dept., Columbia Univ., 1982.

33. U. Bizzari et al., J. de Physique, vol. 44, suppl. 2, p. C1-313, 1983.

34. J.P. Blewett and R. C. Chasman, J. Appl. Phys., vol. 48, p. 2692, 1977.

35. H. Boehmer, J. Munch, M.Z. Caponi, IEEE Trans. Nucl. Sci., vol. NS-26, p. 3830; also in "Free Electron Lasers," S. Martellucci and A.N. Chester eds., (Plenum, 1979), p. 541.

36. M. Borenstein and W.E. Lamb, Jr., Phys. Rev. A, vol. 5, p. 1298, 1972.

37. K. Boyer, C.A. Brau, B.E. Newnam, W.E. Stein, R.W. Warren, J.G. Winston, and L.M. Young, J. de Physique, vol. 44, suppl. 2, p. C1-109, 1983.

38. V. Bratman, Radio Eng. and Electr. Phys., vol. 27, p. 106, 1982.

39. V.L. Bratman, G.G. Denisov, N.S. Ginzburg, and M.I. Petelin, IEEE J. Quant. Elect., vol. QE-19, p. 282, 1983.

40. C.A. Brau and R.K. Cooper, Phys. Quan. Elect., vol. 7, p. 647, 1980.

41. C.A. Brau, IEEE J. Quant. Electr., vol. QE-16, p. 335, 1980.

42. Y. Carmel, V.L. Granatstein, and A. Gover, Phys. Rev. Lett., vol. 51, p. 566, 1983.

43. S.C. Chen and T.C. Marshall, Phys. Rev. Lett., vol. 52, p. 425, 1984.

44. F. Ciocci, G. Dattoli, and A. Renieri, Lett. Nuov. Cim., vol. 34, p. 341, 1982.

45. P. Coleman and C. Enderby, J. Appl. Phys., vol. 31, p. 1695, 1960.

46. W.B. Colson, Phys. Quan. Electr., vol. 5, p. 152, 1977.

47. W.B. Colson and S.K. Ride, Phys. Quan. Electr., vol. 7, p. 377, 1980.

48. W.B. Colson, J. Quan. Elect., vol. QE-17, p. 1417, 1981.

49. W.B. Colson, Phys. Quan. Elect., vol. 8, p. 457, 1982.

50. W.B. Colson and P. Elleaume, Appl. Phys., vol. B29, p. 1, 1982.

51. W.B. Colson and R.A. Freedman, Phys. Rev. A, vol. 27, p. 1399, 1983.

52. W.B. Colson and J.L. Richardson, Phys. Rev. Lett., vol. 50, p. 1050, 1983.

53. G. Dattoli, E. Fiorentino, T. Letardi, A. Marino, and A. Renieri, Trans. Nucl. Sci., vol. NS-28, p. 3133, 1981.

54. D.A.G. Deacon, L.R. Elias, J.M.J. Madey, G.J. Ramian, H.A. Schwettman, and T.L. Smith, Phys. Rev. Lett., vol. 38, p. 892, 1977.

55. D.A.G. Deacon, K.E. Robinson, J.M.J. Madey, C. Bazin, M. Billardon, P. Elleaume, Y. Farge, J.M. Ortega, Y. Petroff, and M.F. Velghe, Opt. Comm., vol. 40, p. 373, 1982.

56. P. Diament, Phys. Rev. A, vol. 23, p. 2537, 1981.

57. A.N. Didenko, A.V. Kozhevnikov, A.F. Medvedev, M.M. Nikitkin, and V.Y. Epp, Sov. Phys. JETP, vol. 49, p. 973, 1979.

58. P. Dobiasch, K.L. Hohla, and P. Meystre, Optics & Lasers in Eng., vol. 4, p. 91, 1983.

59. F. Dolezal, R. Harvey, and C. Parazzoli, IEEE J. Quant. Elect., vol. QE-19, p. 309, 1983.

60. J.N. Eckstein, J.M.J. Madey, K. Robinson, T.I. Smith, S. Benson, D.A.G. Deacon, and R. Taber, Phys. Quan. Elect., vol. 8, p. 49, 1982.

61. J.A. Edighoffer, W.D. Kimura, R.H. Pantell, M.A. Piestrup, D.Y. Wang, Phys. Rev. A, vol. 23, p. 1848, 1981; also J. Quan. Elect., vol. QE-17, p. 1507, 1981.

62. J. Edighoffer, H. Boehmer, M.Z. Caponi, S. Fornaca, J. Munch, G.R. Neil, B. Saur, and C. Shih, J. Quan. Elect., vol. QE-19, p. 316, 1983.

63. J.A. Edighoffer, G.R. Neil, C.E. Hess, T.I. Smith, S.W. Fornaca, and H.A. Schwettman, Phys. Rev. Lett., vol. 52, p. 344, 1984.

64. P. Efthimion and S.P. Schlesinger, Phys. Rev. A, vol. 16, p. 633, 1977.

65. L.R. Elias, W.M. Fairbank, J.M.J. Madey, H.A. Schwettman, and T.I. Smith, Phys. Rev. Lett., vol. 36, p. 710, 1976.

66. L.R. Elias, Phys. Rev. Lett., vol. 42, p. 977; also in "Free Electron Lasers," S. Martellucci and A.N. Chester eds., p. 617 (Plenum, 1979).

67. L.R. Elias and G. Ramian, Phys. Quan. Elect., vol. 9, p. 577, 1982.

68. P. Elleaume, J. de Physique, vol. 44, suppl. 2, p. C1-333, 1983.

69. P. Elleaume and D.A.G. Deacon, Appl. Phys., vol. B32, p. 2621, 1983.

70. R.A. Freedman and W.B. Colson, J. de Physique, vol. 44, suppl. 2, p. C1-369, 1983; also Opt. Commun., vol. 46, p. 37, 1983.

71. H.P. Freund and P. Sprangle, in "Proceedings of Infrared and Millimeter Wave Conference" (IEEE, Dec. 1981).

72. H.P. Freund and A.T. Drobot, Phys. Fluids., vol. 25, p. 736, 1982.

73. H.P. Freund, S. Johnston, and P. Sprangle, J. Quan. Elect., vol. QE-19, p. 322, 1983.

74. H.P. Freund, Phys. Rev. A, vol. 27, p. 1977, 1983.

75. H.P. Freund and S.H. Gold, Phys Rev. Lett. 52, p. 926 (1984).

76. H.P. Freund, P. Sprangle, D. Dillenburg, F.H. da Jornada, R.S. Schneider, and B. Liberman, Phys. Rev. A, vol. 24, p. 1965, 1982.

77. L. Friedland, Phys. Fluids, vol. 23, p. 2376, 1980.

78. M. Friedman and M. Herndon, Phys. Rev. Lett., vol. 28, p. 55, 1972.

79. M. Friedman, D.A. Hammer, W.M. Manheimer, and P. Sprangle, Phys. Rev. Lett., vol. 31, p. 752, 1973.

80. M. Friedman and M. Herndon, Phys. Fluids, vol. 16, p. 1982, 1973.

81. A. Fruchtman and J.L. Hirshfield, Intl. J. IR and MM Waves, vol. 2, p. 905, 1981.

82. A. Fruchtman and L. Friedland, J. Quan. Elect., vol. QE-19, p. 327, 1983; also, A. Fruchtman, J. Appl. Phys., vol. 54, p. 4289, 1983.

83. H. Gamo, J.S. Ostrem, and S.S. Chuang, J. Appl. Phys., vol. 44, p. 2750, 1973.

84. R.M. Gilgenbach, T.C. Marshall, and S.P. Schlesinger, Phys. Fluids, vol. 22, 971, 1979.

85. S.H. Gold, W.M. Black, H.P. Freund, V.L. Granatstein, R.H. Jackson, P.C. Efthimion, and A.K. Kinkead, Phys. Fluids, vol. 26, p. 2683, 1983; also vol. 27, p. 746 (1984).

86. A. Gover and P. Sprangle, J. Quant. Elect., vol. QE-17, p. 1196, 1981.

87. V.L. Granatstein, M. Herndon, R.K. Parker, and S.P. Schlesinger, IEEE Trans., vol. MTT-22, p. 1000, 1974.

88. V.L. Granatstein, S.P. Schlesinger, M. Herndon, R.K. Parker, J.A. Pasour, Appl. Phys. Lett., vol. 30, p. 384, 1977.

89. A. Grossman, "A New Millimeter FEL Using the Cyclotron-Undulator Inter- actions of Relativistic Spiralling Electrons," Thesis, Columbia Univ., Dept. of Physics, 1982.

90. A. Grossman, T.C. Marshall, and S.P. Schlesinger, Phys. Fluids, vol. 26, p. 337, 1983.

91. W.M. Grossman, T.L. Churchill, et al., in "Free Electron Generators of

Coherent Radiation," #453, p. 52 (SPIE, 1983), and Appl. Phys. Lett., vol. 43, p. 745 (1983).

92. A. Grossman and T.C. Marshall, J. Quan. Elect., vol. QE-19, p. 334, 1983.

93. W.M. Grossman and D.C. Quimby, in "Free Electron Generators of Coherent Radiation," #453, p. 86 (SPIE, 1983).

94. K. Halbach, Nucl. Inst. and Meth., vol. 169, p. 1, 1980; vol. 187, p. 109, 1981.

95. A. Hasegawa, K. Mima, P. Sprangle, H.H. Szu, and V.L. Granatstein, Appl. Phys. Lett., vol. 29, p. 542, 1976.

96. A. Hasegawa, Bell. Syst. Tech. J., vol. 57, p. 3069, 1978.

97. J.L. Hirshfield, K.R. Chu, and S. Kainer, Appl. Phys. Lett., vol. 33, p. 847, 1978.

98. F.A. Hopf, P. Meystre, M.O. Scully, and J.F. Seely, Phys. Rev. Lett., vol. 35, p. 511, 1975.

99. L.F. Ibanez and S. Johnston, J. Quan. Elect., vol. QE-19, p. 339, 1983.

100. R.H. Jackson, S.H. Gold, R.K. Parker, H.P. Freund, P. Efthimion, V.L. Granatstein, M. Herndon, A.K. Kinkead, J.E. Kosakowski, and T.J. Kwan, J. Quan. Elect., vol. QE-19, p. 346, 1983.

101. K.D. Jacobs, R.E. Shefer, and G. Bekefi, Appl. Phys. Lett., vol. 37, p. 583, 1980.

102. S. Johnston, Phys. Quan. Elect., vol. 8, p. 325, 1982.

103. C.A. Kapetanakos, P. Sprangle, D.P. Chernin, S.J. Marsh, and I. Haber, Phys. Fluids, vol. 26, p. 1634, 1983.

104. P.L. Kapitza and P.A.M. Dirac, Proc. Cambr. Phil. Soc., vol. 29, p 297, 1933.

105. B.M. Kincaid, J. Appl. Phys., vol. 48, p. 2684, 1977.

106. F.K. Kneubuhl, in "Proceedings, Conference on Lasers '79," p. 812 (Society for Optical and Quantum Electronics, 1980).

107. N.M. Kroll and W.A. McMullin, Phys. Rev. A, vol. 17, p. 300, 1978.

108. N.M. Kroll, P.L. Morton, M.N. Rosenbluth, J.N. Eckstein, and J.M.J. Madey, J. Quant. Elect., vol. QE-17, p. 1496, 1981.

109. N. Kroll, Phys. Quan. Elect., vol. 8, p. 281, 1982.

110. N.M. Kroll and M.N. Rosenbluth, J. de Physique, vol. 44, suppl. 2, p. C1-85, 1983.

111. A.B. Kukushkin, Sov. J. Plasma Phys., vol. 7, p. 63, 1981.

112. T. Kwan, J.M. Dawson, and A.T. Lin, Phys. Fluids, vol. 20, p. 581, 1977.

113. T.J.T. Kwan, Thesis, report PPG-354, UCLA, 1978.

114. T.J.T. Kwan, J. Quan. Elect., vol. QE-17, p. 1394, 1981.

115. W. Lamb, Phys. Rev., vol. 134, p. 1429, 1964.

116. J.D. Lawson, IEEE Trans. Nucl. Sci., vol. NS-26, p. 4217, 1979.

117. R.P. Leavitt, D.E. Wortman, and H. Dropkin, J. Quant. Elect., vol. QE-17, pp. 1333, 1341; 1981.

118. J.E. Leiss, N.J. Norris, and M.A. Wilson, Part. Accel., vol. 10, p. 223, 1980.

119. P.C. Liewer, A.T. Lin, and J.M. Dawson, Phys. Rev. A, vol. 23, p. 1251, 1981.

120. A.T. Lin and J.M. Dawson, Phys. Rev. Lett., vol. 42, p. 1670, 1979.

121. A.T. Lin and J.M. Dawson, Phys. Quan. Elect., vol. 7, p. 555, 1980.

122. A.T. Lin, Phys. Quan. Elect., vol. 9, p. 867, 1982.

123. A.T. Lin, W.W. Chang, C.C. Lin, Phys. Flds., vol. 27, p. 1054 (1984).

124. W.H. Louisell, J.F. Lam, and D.A. Copeland, Phys. Rev. A, vol. 18, p. 655 (1978).

125. J.M.J. Madey, J. Appl. Phys., vol. 42, p. 1906 (1971).

126. J.M.J. Madey, H.A. Schwettman, and W.M. Fairbank, Trans. Nucl. Sci., vol. NS-20, p. 980, 1973.

127. J.M.J. Madey, Nuovo Cim., vol. 50B, p. 64, 1979.

128. J.M.J. Madey, J. de Physique, vol. 44, suppl. 2, p. C1-169, 1983.

129. W.M. Manheimer and E. Ott, Phys. Fluids, vol. 17, p. 1413, 1974.

130. S.A. Mani, D.A. Korff, and J. Blimmel, Phys. Quan. Elect., vol. 9, p. 557, 1982.

131. T.C. Marshall, S. Talmadge, and P. Efthimion, Appl. Phys. Lett., vol. 31, p. 320, 1977.

132. D.B. McDermott, T.C. Marshall, S.P. Schlesinger, R.K. Parker, V.L. Granatstein, Phys. Rev. Lett., vol. 41, p. 1368, 1978.

133. D.B. McDermott and T.C. Marshall, Phys. Quan. Elect., vol. 7, p. 509, 1980.

134. W.A. McMullin and G. Bekefi, Appl. Phys. Lett., vol. 39, p. 845, 1981.

135. W.A. McMullin and G. Bekefi, Phys. Rev. A, vol. 25, p. 1826, 1982.

136. W.A. McMullin, R.C. Davidson, and G.L. Johnston, report PFC/JA 83-10 MIT, 1983.

137. W.A. McMullin and R.C. Davidson, Phys. Rev. A, vol. 25, p. 3130, 1982.

138. G. Moore, J. Gea-Benacloche, et al., in "Free Electron Generators of Coherent Radiation," #453, p. 393 (SPIE, 1983).

139. P.L. Morton, Phys. Quan. Elect., vol. 8, p. 1, 1982.

140. H. Motz, J. Appl. Phys., vol. 22, p. 527, 1951.

141. H. Motz, W. Thon, and R.N. Whitehorst, J. Appl. Phys., vol. 24, p. 826, 1953.

142. J.A. Nation, Appl. Phys. Lett., vol. 17, p. 491, 1970; see also Phys. Rev. Lett., vol. 33, p. 1278, 1974.

143. V.K. Neil, Jason Report JSR-79-10, SRI International, 1979.

144. B.E. Newnam, K. Boyer, C.A. Brau, and R.W. Warren, in "Free Electron Generators of Coherent Radiation," #453, p. 118 (SPIE, 1983).

145. J.H. Nuckolls, Physics Today, Sept. 1982, p. 25.

146. E. Ott and W.M. Manheimer, IEEE Trans. Plasma Sci., vol. PS-13, p. 1, 1975.

147. R.B. Palmer, J. Appl. Phys., vol. 43, p. 3014, 1972.

148. R.H. Pantell, G. Soncini, and H.E. Puthoff, J. Quan. Elect., vol. 4, p. 905, 1968.

149. R.H. Pantell and M.J. Alguard, J. Appl. Phys., vol. 50, p. 798, 1983.

150. S.Y. Park, J.M. Baird, R.A. Smith, and J.L. Hirshfield, J. Appl. Phys., vol. 53, p. 1320, 1982.

151. R.K. Parker, R.H. Jackson, S.H. Gold, H.P. Freund, V.L. Granatstein, P.C. Efthimion, M. Herndon, and A.K. Kinkead, Phys. Rev. Lett., vol. 48, p. 238, 1982.

152. J.A. Pasour, C.W. Roberson, and F. Mako, J. Appl. Phys., vol. 53, p. 7174, 1982.

153. C. Pellegrini, Trans. Nucl. Sci., vol. NS-26, p. 3791; also, in "Free Electron Lasers," S. Martellucci and A.N. Chester, eds., p. 91 (Plenum, 1979).

154. C. Pellegrini, in "AIP Conference Proceedings #91," P.J. Channel ed. (Amer. Inst. of Physics, New York, 1982).

155. R.M. Phillips, Trans. IRE El. Dev., vol. 7, p. 231, 1960.

156. D. Prosnitz, A. Szoke, and V.K. Neil, Phys. Rev. A, vol. 24, p. 1436, 1981.

157. G. Providakes and J.A. Nation, J. Appl. Phys., vol. 50, p. 3026, 1979.

158. D. Quimby and J.M. Slater, in "Free Electron Generators of Coherent Radiation," #453, p. 92 (SPIE, 1983).

159. M. Reiser, Phys. Fluids, vol. 20, p. 477, 1977.

160. A. Renieri, Frascati, CNEN Report 77/33, 1977.

161. A. Renieri, Nuovo Cim., vol. 53B, p. 160, 1979.

162. S.K. Ride and W.B. Colson, Appl. Phys. vol. 20, p. 41, 1979.

163. C.W. Roberson, J.A. Pasour, F. Mako, R. Lucey, and P. Sprangle, Infrared and MM Waves, vol. 10, p. 361, 1983.

164. K.E. Robinson, D.A.G. Deacon, M.F. Velghe, and J.M.J. Madey, J. Quan. Elect., vol. QE-19, p. 365, 1983.

165. F.S. Rusin and G.D. Bogomolov, JETP Lett., vol. 4, p. 160, 1966.

166. S.B. Segall et al., "Review of Two-Stage FEL Research at KMS Fusion," report #KMSF-U1307, 1983; also in "Free Electron Generators of Coherent Radiation," #453, p. 178 (SPIE, 1983).

167. E.D. Shaw and C.K.N. Patel, Phys. Quan. Elect., vol. 7, p. 665; vol. 9, p. 671, 1980.

168. E.D. Shaw and C.K.N. Patel, Phys. Rev. Lett., vol. 46, p. 332, 1981.

169. E.D. Shaw and C.K.N. Patel, in "Proceedings of the International Conference LASERS '80," pp. 53–60 (1981).

170. E.D. Shaw, R.M. Emanuelson, and G.A. Herbster, J. de Physique, vol. 44, suppl. 2, p. C1-153, 1983.

171. R.L. Sheffield, M.D. Montgomery, J.V. Parker, K.B. Riepe, and S. Singer, J. Appl. Phys, vol. 53, p. 5408, 1982.

172. R.E. Shefer and G. Bekefi, Phys. Quan. Elect., vol. 9, p. 703, 1982.

173. R.E. Shefer, Y.Z. Yin, and G. Bekefi, J. Appl. Phys., vol. 54, p. 6154, 1983.

174. C.C. Shih and M.Z. Caponi, J. Quan. Elect., vol. QE-19, p. 369, 1983.

175. C.C. Shih, G.R. Neil, J. Munch, S. Fornaca, J.A. Edighoffer, M.Z. Caponi, and H.E. Boehmer, J. de Physique, vol. 44, suppl. 2, p. C1-115, 1983

176. J.M. Slater, J. Quan. Elect., vol. QE-17, p. 1476, 1981.

177. J.M. Slater, J. Adamski, T.L. Churchill, L.Y. Nelson, and R.E. Carter, J. Quan. Elect., vol. QE-19, p. 374, 1983.

178. M.L. Sloan, and A.A. Davis, Phys. Fluids, vol. 25, p. 2337, 1982.

179. S.J. Smith and E.M. Purcell, Phys. Rev., vol. 91, p. 1069, 1953.

180. T.I. Smith, J.M.J. Madey, L.R. Elias, and D.A.G. Deacon, J. Appl. Phys., vol. 50, p. 4580, 1979.

181. T.I. Smith, Phys. Quan. Elect., vol. 8, p. 77, 1982.

182. P. Sprangle, V.L. Granatstein, and L. Baker, Phys. Rev. A, vol. 12, p. 1697, 1975.

183. P. Sprangle and C.M. Tang, Trans. Nucl. Sci., vol. NS-28, p. 3346, 1981.

184. P. Sprangle and A.T. Drobot, J. Appl. Phys., vol. 50, p. 2652, 1979.

185. P. Sprangle, C.M. Tang, and W.M. Manheimer, Phys. Rev. Lett., vol. 43, p. 1932, 1979.

186. P. Sprangle, C.M. Tang, and I. Bernstein, Phys. Rev. Lett., vol. 50, p. 1775, 1983; also, Phys. Rev. A, vol. 28, p. 2300, 1983.

187. V.P. Sukhatme and P.W. Wolff, J. Appl. Phys., vol. 44, p. 2331, 1973.

188. H. Takeda and S.B. Segall, in "Free Electron Generators of Coherent Radiation," #453, p. 196 (SPIE, 1983).

189. S. Talmadge, T.C. Marshall, and S.P. Schlesinger, Phys. Fluids, vol. 20, p. 974, 1977.

190. C.M. Tang and P. Sprangle, J. Appl. Phys., vol. 52, p. 3148, 1981.

191. C.M. Tang and P. Sprangle, Phys. Quan. Elect., vol. 9, p. 627, 1982.

192. H.S. Uhm and R.C. Davidson, Phys. Fluids, vol. 24, p. 1541, 1981; also Phys. Fluids, vol. 24, p. 2348, 1981.

193. S. von Laven, J. Branscum, J. Golub, R. Layman, and J. Walsh, Appl. Phys. Lett., vol. 41, p. 408, 1982.

194. J.E. Walsh, T.C. Marshall, and S.P. Schlesinger, Phys. Fluids, vol. 20, p. 709, 1977.

195. J.E. Walsh, Phys. Quan. Elect., vol. 7, p. 255, 1980.

196. J. Walsh, Adv. Elect. and Elect. Phys., C. Marton ed., vol. 58, p. 271, (Academic Press, 1982).

197. R.W. Warren, B.E. Newnam, J.Q. Winston, W.E. Stern, L.M. Young, and C.A. Brau, J. Quan. Elect., vol. QE-19, p. 391, 1983.

198. H. Wiedemann, J. de Physique, vol. 44, suppl. 2, p. C1-201, 1983.

199. T.J. Orzechowski, et al., in "Free Electron Generators of Coherent Radiation," #453, p. 65 (SPIE, 1983).

200. V.A. Zhuravlev and G.D. Petrov, Sov. J. Plasma Phys., vol. 5, p. 3, 1979.

References Added in Proof

201. A.G. Fox and T. Li, Bell Syst. Tech. J., vol. 40, p. 453 (1961).

202. H. Motz and M. Nakamura, Annals of Physics, vol. 7, p. 84 (1959).

203. K. Landecker, Phys. Rev., vol. 86, p. 852 (1952).

204. D.B. McDermott, in "A Collective Free Electron Laser," Thesis, Columbia University, 1979.

205. F.A. Hopf, P. Meystre, M.O. Scully, and W.H. Louisell, Phys. Rev. Lett., vol. 37, p. 1215 (1976).

206. S. Benson, D.A.G. Deacon, J.N. Eckstein, J.M.J. Madey, K. Robinson, T.I. Smith, and R. Taber, Phys. Rev. Lett., vol. 48, p. 235 (1982).

INDEX

Index

187